We Have Never
Been M

D0707753

We Have Never Been Modern

Bruno Latour

translated by Catherine Porter

Harvard University Press
Cambridge, Massachusetts

Originally published as *Nous n'avons jamais été modernes:
Essais d'anthropologie symmétrique.*

Copyright © 1991 La Découverte

Library of Congress Cataloging-in-Publication Data

Latour, Bruno.
    [Nous n'avons jamais été modernes. English]
    We have never been modern / Bruno Latour : translated by
Catherine Porter.
      p. cm.
    Translation of: Nous n'avons jamais été moderns.
    Includes bibliographical references and index.
    ISBN 0–674–94838–6. — ISBN 0–674–94839–4 (pbk.)
    1. Science–Social aspects. 2. Technology–Social aspects.
3. Science–Philosophy. 4. Science–History. I. Title.
Q175.5.L3513   1993
303.48′3–dc20   93–15226
                                       CIP

For Charis and Adrian

CONTENTS

ACKNOWLEDGEMENTS

In many places the English text differs from the French. I have modified the figures, added section 3.2 and qualified or clarified the argument without modifying its main structure. I have abstained from giving empirical examples in order to retain the speculative – and, I am afraid, very Gallic! – character of this essay. Many case studies, including several by myself, will be found in the bibliography. Having written several empirical books, I am trying here to bring the emerging field of science studies to the attention of the literate public through the philosophy associated with this domain.

Many people have tried to make this essay less unreasonable. Among them I especially thank Luc Boltanski, Francis Chateauraynaud, Elizabeth Claverie, Gerard de Vries, François Gèze and Isabelle Stengers.

I thank Harry Collins, Ernan McMullin, Jim Griesemer, Michel Izard, Clifford Geertz and Peter Galison for allowing me to present the arguments of this essay in their seminars.

Parts of Chapter 2 have been published in 'Postmodern? No, simply amodern: steps towards an anthropology of science. An essay review', *Studies in the History and Philosophy of Science* 21: (1990) 145–71. Some of the arguments in Chapter 3 have appeared in a different form in 'One more turn after the social turn: easing science studies into the nonmodern world', in E. McMullin, ed., *The Social Dimensions of Science*. Notre Dame: Notre Dame University Press, 1992, pp. 272–92.

I

□

CRISIS

1.1 The Proliferation of Hybrids

On page four of my daily newspaper, I learn that the measurements taken above the Antarctic are not good this year: the hole in the ozone layer is growing ominously larger. Reading on, I turn from upper-atmosphere chemists to Chief Executive Officers of Atochem and Monsanto, companies that are modifying their assembly lines in order to replace the innocent chlorofluorocarbons, accused of crimes against the ecosphere. A few paragraphs later, I come across heads of state of major industrialized countries who are getting involved with chemistry, refrigerators, aerosols and inert gases. But at the end of the article, I discover that the meteorologists don't agree with the chemists; they're talking about cyclical fluctuations unrelated to human activity. So now the industrialists don't know what to do. The heads of state are also holding back. Should we wait? Is it already too late? Toward the bottom of the page, Third World countries and ecologists add their grain of salt and talk about international treaties, moratoriums, the rights of future generations, and the right to development.

The same article mixes together chemical reactions and political reactions. A single thread links the most esoteric sciences and the most sordid politics, the most distant sky and some factory in the Lyon suburbs, dangers on a global scale and the impending local elections or the next board meeting. The horizons, the stakes, the time frames, the actors – none of these is commensurable, yet there they are, caught up in the same story.

On page six, I learn that the Paris AIDS virus contaminated the culture medium in Professor Gallo's laboratory; that Mr Chirac and Mr Reagan had, however, solemnly sworn not to go back over the history of that

I

discovery; that the chemical industry is not moving fast enough to market medications which militant patient organizations are vocally demanding; that the epidemic is spreading in sub-Saharan Africa. Once again, heads of state, chemists, biologists, desperate patients and industrialists find themselves caught up in a single uncertain story mixing biology and society.

On page eight, there is a story about computers and chips controlled by the Japanese; on page nine, about the right to keep frozen embryos; on page ten, about a forest burning, its columns of smoke carrying off rare species that some naturalists would like to protect; on page eleven, there are whales wearing collars fitted with radio tracking devices; also on page eleven, there is a slag heap in northern France, a symbol of the exploitation of workers, that has just been classified as an ecological preserve because of the rare flora it has been fostering! On page twelve, the Pope, French bishops, Monsanto, the Fallopian tubes, and Texas fundamentalists gather in a strange cohort around a single contraceptive. On page fourteen, the number of lines on high-definition television bring together Mr Delors, Thomson, the EEC, commissions on standardization, the Japanese again, and television film producers. Change the screen standard by a few lines, and billions of francs, millions of television sets, thousands of hours of film, hundreds of engineers and dozens of CEOs go down the drain.

Fortunately, the paper includes a few restful pages that deal purely with politics (a meeting of the Radical Party), and there is also the literary supplement in which novelists delight in the adventures of a few narcissistic egos ('I love you . . . you don't'). We would be dizzy without these soothing features. For the others are multiplying, those hybrid articles that sketch out imbroglios of science, politics, economy, law, religion, technology, fiction. If reading the daily paper is modern man's form of prayer, then it is a very strange man indeed who is doing the praying today while reading about these mixed-up affairs. All of culture and all of nature get churned up again every day.

Yet no one seems to find this troubling. Headings like Economy, Politics, Science, Books, Culture, Religion and Local Events remain in place as if there were nothing odd going on. The smallest AIDS virus takes you from sex to the unconscious, then to Africa, tissue cultures, DNA and San Francisco, but the analysts, thinkers, journalists and decision-makers will slice the delicate network traced by the virus for you into tidy compartments where you will find only science, only economy, only social phenomena, only local news, only sentiment, only sex. Press the most innocent aerosol button and you'll be heading for the Antarctic, and from there to the University of California at Irvine, the mountain ranges of Lyon, the chemistry of inert gases, and then maybe to the

United Nations, but this fragile thread will be broken into as many segments as there are pure disciplines. By all means, they seem to say, let us not mix up knowledge, interest, justice and power. Let us not mix up heaven and earth, the global stage and the local scene, the human and the nonhuman. 'But these imbroglios do the mixing,' you'll say, 'they weave our world together!' 'Act as if they didn't exist,' the analysts reply. They have cut the Gordian knot with a well-honed sword. The shaft is broken: on the left, they have put knowledge of things; on the right, power and human politics.

1.2 Retying the Gordian Knot

For twenty years or so, my friends and I have been studying these strange situations that the intellectual culture in which we live does not know how to categorize. For lack of better terms, we call ourselves sociologists, historians, economists, political scientists, philosophers or anthropologists. But to these venerable disciplinary labels we always add a qualifier: 'of science and technology'. 'Science studies', as Anglo-Americans call it, or 'science, technology and society'. Whatever label we use, we are always attempting to retie the Gordian knot by crisscrossing, as often as we have to, the divide that separates exact knowledge and the exercise of power – let us say nature and culture. Hybrids ourselves, installed lopsidedly within scientific institutions, half engineers and half philosophers, 'tiers instruits' (Serres, 1991) without having sought the role, we have chosen to follow the imbroglios wherever they take us. To shuttle back and forth, we rely on the notion of translation, or network. More supple than the notion of system, more historical than the notion of structure, more empirical than the notion of complexity, the idea of network is the Ariadne's thread of these interwoven stories.

Yet our work remains incomprehensible, because it is segmented into three components corresponding to our critics' habitual categories. They turn it into nature, politics or discourse.

When Donald MacKenzie describes the inertial guidance system of intercontinental missiles (MacKenzie, 1990); when Michel Callon describes fuel cell electrodes (Callon, 1989); when Thomas Hughes describes the filament of Edison's incandescent lamp (Hughes, 1983); when I describe the anthrax bacterium modified by Louis Pasteur (Latour, 1988b) or Roger Guillemin's brain peptides (Latour and Woolgar, [1979] 1986), the critics imagine that we are talking about science and technology. Since these are marginal topics, or at best manifestations of pure instrumental and calculating thought, people who are interested in politics or in souls feel justified in paying no attention.

Yet this research does not deal with nature or knowledge, with things-in-themselves, but with the way all these things are tied to our collectives and to subjects. We are talking not about instrumental thought but about the very substance of our societies. MacKenzie mobilizes the entire American Navy, and even Congress, to talk about his inertial guidance system; Callon mobilizes the French electric utility (EDF) and Renault as well as great chunks of French energy policy to grapple with changes in ions at the tip of an electrode in the depth of a laboratory; Hughes reconstructs all America around the incandescent filament of Edison's lamp; the whole of French society comes into view if one tugs on Pasteur's bacteria; and it becomes impossible to understand brain peptides without hooking them up with a scientific community, instruments, practices – all impedimenta that bear very little resemblance to rules of method, theories and neurons.

'But then surely you're talking about politics? You're simply reducing scientific truth to mere political interests, and technical efficiency to mere strategical manœuvres?' Here is the second misunderstanding. If the facts do not occupy the simultaneously marginal and sacred place our worship has reserved for them, then it seems that they are immediately reduced to pure local contingency and sterile machinations. Yet science studies are talking not about the social contexts and the interests of power, but about their involvement with collectives and objects. The Navy's organization is profoundly modified by the way its offices are allied with its bombs; EDF and Renault take on a completely different look depending on whether they invest in fuel cells or the internal combustion engine; America before electricity and America after are two different places; the social context of the nineteenth century is altered according to whether it is made up of wretched souls or poor people infected by microbes; as for the unconscious subjects stretched out on the analyst's couch, we picture them differently depending on whether their dry brain is discharging neurotransmitters or their moist brain is secreting hormones. None of our studies can reutilize what the sociologists, the psychologists or the economists tell us about the social context or about the subject in order to apply them to the hard sciences – and this is why I will use the word 'collective' to describe the association of humans and nonhumans and 'society' to designate one part only of our collectives, the divide invented by the social sciences. The context and the technical content turn out to be redefined every time. Just as epistemologists no longer recognize in the collectivized things we offer them the ideas, concepts or theories of their childhood, so the human sciences cannot be expected to recognize the power games of their militant adolescence in these collectives full of things we are lining up. The delicate networks traced by Ariadne's little hand remain more invisible than spiderwebs.

'But if you are not talking about things-in-themselves or about humans-among-themselves, then you must be talking just about discourse, representation, language, texts, rhetorics.' This is the third misunderstanding. It is true that those who bracket off the external referent – the nature of things – and the speaker – the pragmatic or social context – can talk only about meaning effects and language games. Yet when MacKenzie examines the evolution of inertial guidance systems, he is talking about arrangements that can kill us all; when Callon follows a trail set forth in scientific articles, he is talking about industrial strategy as well as rhetoric (Callon *et al.*, 1986); when Hughes analyzes Edison's notebooks, the internal world of Menlo Park is about to become the external world of all America (Hughes, 1983). When I describe Pasteur's domestication of microbes, I am mobilizing nineteenth-century society, not just the semiotics of a great man's texts; when I describe the invention-discovery of brain peptides, I am really talking about the peptides themselves, not simply their representation in Professor Guillemin's laboratory. Yet rhetoric, textual strategies, writing, staging, semiotics – all these are really at stake, but in a new form that has a simultaneous impact on the nature of things and on the social context, while it is not reducible to the one or the other.

Our intellectual life is out of kilter. Epistemology, the social sciences, the sciences of texts – all have their privileged vantage point, provided that they remain separate. If the creatures we are pursuing cross all three spaces, we are no longer understood. Offer the established disciplines some fine sociotechnological network, some lovely translations, and the first group will extract our concepts and pull out all the roots that might connect them to society or to rhetoric; the second group will erase the social and political dimensions, and purify our network of any object; the third group, finally, will retain our discourse and rhetoric but purge our work of any undue adherence to reality – *horresco referens* – or to power plays. In the eyes of our critics the ozone hole above our heads, the moral law in our hearts, the autonomous text, may each be of interest, but only separately. That a delicate shuttle should have woven together the heavens, industry, texts, souls and moral law – this remains uncanny, unthinkable, unseemly.

1.3 The Crisis of the Critical Stance

The critics have developed three distinct approaches to talking about our world: naturalization, socialization and deconstruction. Let us use E.O. Wilson, Pierre Bourdieu, and Jacques Derrida – a bit unfairly – as emblematic figures of these three tacks. When the first speaks of naturalized phenomena, then societies, subjects, and all forms of discourse vanish. When the second speaks of fields of power, then science, technology, texts, and the contents of activities disappear. When the third speaks of truth effects, then to believe in the real existence of brain neurons or power plays would betray enormous naiveté. Each of these forms of criticism is powerful in itself but impossible to combine with the other two. Can anyone imagine a study that would treat the ozone hole as simultaneously naturalized, sociologized and deconstructed? A study in which the nature of the phenomena might be firmly established and the strategies of power predictable, but nothing would be at stake but meaning effects that project the pitiful illusions of a nature and a speaker? Such a patchwork would be grotesque. Our intellectual life remains recognizable as long as epistemologists, sociologists and deconstructionists remain at arm's length, the critique of each group feeding on the weaknesses of the other two. We may glorify the sciences, play power games or make fun of the belief in a reality, but we must not mix these three caustic acids.

Now we cannot have it both ways. Either the networks my colleagues in science studies and I have traced do not really exist, and the critics are quite right to marginalize them or segment them into three distinct sets: facts, power and discourse; or the networks are as we have described them, and they do cross the borders of the great fiefdoms of criticism: they are neither objective nor social, nor are they effects of discourse, even though they are real, and collective, and discursive. Either we have to disappear, we bearers of bad news, or criticism itself has to face a crisis because of these networks it cannot swallow. Yes, the scientific facts are indeed constructed, but they cannot be reduced to the social dimension because this dimension is populated by objects mobilized to construct it. Yes, those objects are real but they look so much like social actors that they cannot be reduced to the reality 'out there' invented by the philosophers of science. The agent of this double construction – science with society and society with science – emerges out of a set of practices that the notion of deconstruction grasps as badly as possible. The ozone hole is too social and too narrated to be truly natural; the strategy of industrial firms and heads of state is too full of chemical reactions to be reduced to power and interest; the discourse of the ecosphere is too real and too social to boil down to meaning effects. Is it our fault if the networks are *simultaneously real, like nature, narrated, like discourse, and collective, like society*? Are we to pursue them while abandoning all the resources of criticism, or are we to abandon them while endorsing the common sense of the critical tripartition? The tiny networks we have unfolded are torn apart like the Kurds by the Iranians, the Iraqis and the Turks; once night has fallen, they slip across borders to get married, and they dream of a common homeland that would be carved out of the three countries which have divided them up.

This would be a hopeless dilemma had anthropology not accustomed us to dealing calmly and straightforwardly with the seamless fabric of what I shall call 'nature-culture', since it is a bit more and a bit less than a culture (see Section 4.5). Once she has been sent into the field, even the most rationalist ethnographer is perfectly capable of bringing together in a single monograph the myths, ethnosciences, genealogies, political forms, techniques, religions, epics and rites of the people she is studying. Send her off to study the Arapesh or the Achuar, the Koreans or the Chinese, and you will get a single narrative that weaves together the way people regard the heavens and their ancestors, the way they build houses and the way they grow yams or manioc or rice, the way they construct their government and their cosmology. In works produced by anthropologists abroad, you will not find a single trait that is not simultaneously real, social and narrated.

If the analyst is subtle, she will retrace networks that look exactly like the sociotechnical imbroglios that we outline when we pursue microbes, missiles or fuel cells in our own Western societies. We too are afraid that the sky is falling. We too associate the tiny gesture of releasing an aerosol spray with taboos pertaining to the heavens. We too have to take laws, power and morality into account in order to understand what our sciences are telling us about the chemistry of the upper atmosphere.

Yes, but we are not savages; no anthropologist studies us that way, and it is impossible to do with our own culture – or should I say nature-culture? – what can be done elsewhere, with others. Why? Because we are modern. Our fabric is no longer seamless. Analytic continuity has become impossible. For traditional anthropologists, there is not – there cannot be, there should not be – an anthropology of the modern world (Latour, 1988a). The ethnosciences can be connected in part to society and to discourse (Conklin, 1983); science cannot. It is even because they remain incapable of studying themselves in this way that ethnographers are so critical, and so distant, when they go off to the tropics to study others. The critical tripartition protects them because it authorizes them to reestablish continuity among the communities of the premoderns. It is only because they separate at home that ethnographers make so bold as to unify abroad.

The formulation of the dilemma is now modified. Either it is impossible to do an anthropological analysis of the modern world – and then there is every reason to ignore those voices claiming to have a homeland to offer the sociotechnological networks; or it is possible to do an anthropological analysis of the modern world – but then the very definition of the modern world has to be altered. We pass from a limited problem – why do the networks remain elusive? Why are science studies ignored? – to a broader and more classical problem: what does it mean to be modern? When we dig beneath the surface of our elders' surprise at the networks that – as we see it – weave our world, we discover the anthropological roots of that lack of understanding. Fortunately, we are being assisted by some major events that are burying the old critical mole in its own burrows. If the modern world in its turn is becoming susceptible to anthropological treatment, this is because something has happened to it. Ever since Madame de Guermantes's salon, we have known that it took a cataclysm like the Great War for intellectual culture to change its habits slightly and open its doors to the upstarts who had been beyond the pale before.

1.4 1989: The Year of Miracles

All dates are conventional, but 1989 is a little less so than some. For everyone today, the fall of the Berlin Wall symbolizes the fall of socialism. 'The triumph of liberalism, of capitalism, of the Western democracies over the vain hopes of Marxism': such is the victory communiqué issued by those who escaped Leninism by the skin of their teeth. While seeking to abolish man's exploitation of man, socialism had magnified that exploitation immeasurably. It is a strange dialectic that brings the exploiter back to life and buries the gravedigger, having given the world lessons in large-scale civil war. The repressed returns, and with a vengeance: the exploited people, in whose name the avant-garde of the proletariat had reigned, becomes a people once again; the voracious elites that were to have been dispensed with return at full strength to take up their old work of exploitation in banks, businesses and factories. The liberal West can hardly contain itself for joy. It has won the Cold War.

But the triumph is short-lived. In Paris, London and Amsterdam, this same glorious year 1989 witnesses the first conferences on the global state of the planet: for some observers they symbolize the end of capitalism and its vain hopes of unlimited conquest and total dominion over nature. By seeking to reorient man's exploitation of man toward an exploitation of nature by man, capitalism magnified both beyond measure. The repressed returns, and with a vengeance: the multitudes that were supposed to be saved from death fall back into poverty by the hundreds of millions; nature, over which we were supposed to gain absolute mastery, dominates us in an equally global fashion, and threatens us all. It is a strange dialectic that turns the slave into man's owner and master, and that suddenly informs us that we have invented ecocides as well as large-scale famine.

The perfect symmetry between the dismantling of the wall of shame and the end of limitless Nature is invisible only to the rich Western democracies. The various manifestations of socialism destroyed both their peoples and their ecosystems, whereas the powers of the North and the West have been able to save their peoples and some of their countrysides by destroying the rest of the world and reducing its peoples to abject poverty. Hence a double tragedy: the former socialist societies think they can solve both their problems by imitating the West; the West thinks it has escaped both problems and believes it has lessons for others even as it leaves the Earth and its people to die. The West thinks it is the sole possessor of the clever trick that will allow it to keep on winning indefinitely, whereas it has perhaps already lost everything.

After seeing the best of intentions go doubly awry, we moderns from the Western world seem to have lost some of our self-confidence. Should we *not* have tried to put an end to man's exploitation of man? Should we *not* have tried to become nature's masters and owners? Our noblest virtues were enlisted in the service of these twin missions, one in the political arena and the other in the domain of science and technology. Yet we are prepared to look back on our enthusiastic and right-thinking youth as young Germans look to their greying parents and ask: 'What criminal orders did we follow?' 'Will we say that we didn't know?'

This doubt about the well-foundedness of the best of intentions pushes some of us to become reactionaries, in one of two ways. We must no longer try to put an end to man's domination of man, say some; we must no longer try to dominate nature, say others. Let us be resolutely antimodern, they all say.

From a different vantage point, the vague expression of postmodernism aptly sums up the incomplete scepticism of those who reject both reactions. Unable to believe the dual promises of socialism and 'naturalism', the postmoderns are also careful not to reject them totally. They remain suspended between belief and doubt, waiting for the end of the millennium.

Finally, those who reject ecological obscurantism or antisocialist obscurantism, and are unable to settle for the scepticism of the postmoderns, decide to carry on as if nothing had changed: they intend to remain resolutely modern. They continue to believe in the promises of the sciences, or in those of emancipation, or both. Yet their faith in modernization no longer rings quite true in art, or economics, or politics, or science, or technology. In art galleries and concert halls, along the façades of apartment buildings and inside international organizations, you can feel that the heart is gone. The will to be modern seems hesitant, sometimes even outmoded.

Whether we are antimodern, modern or postmodern, we are all called
into question by the double debacle of the miraculous year 1989. But we
take up the threads of thought if we consider the year precisely to be a
double debacle, two lessons whose admirable symmetry allows us to look
at our whole past in a new light.

And what if we had never been modern? Comparative anthropology
would then be possible. The networks would have a place of their own.

1.5 What Does it Mean To Be a Modern?

Modernity comes in as many versions as there are thinkers or journalists,
yet all its definitions point, in one way or another, to the passage of time.
The adjective 'modern' designates a new regime, an acceleration, a rupture,
a revolution in time. When the word 'modern', 'modernization', or
'modernity' appears, we are defining, by contrast, an archaic and stable
past. Furthermore, the word is always being thrown into the middle of a
fight, in a quarrel where there are winners and losers, Ancients and
Moderns. 'Modern' is thus doubly asymmetrical: it designates a break in
the regular passage of time, and it designates a combat in which there are
victors and vanquished. If so many of our contemporaries are reluctant
to use this adjective today, if we qualify it with prepositions, it is because
we feel less confident in our ability to maintain that double asymmetry:
we can no longer point to time's irreversible arrow, nor can we award a
prize to the winners. In the countless quarrels between Ancients and
Moderns, the former come out winners as often as the latter now, and
nothing allows us to say whether revolutions finish off the old regimes or
bring them to fruition. Hence the scepticism that is oddly called
'post'modern even though it does not know whether or not it is capable
of taking over from the Moderns.

To go back a few steps: we have to rethink the definition of modernity,
interpret the symptom of postmodernity, and understand why we are no
longer committed heart and soul to the double task of domination and
emancipation. To make a place for the networks of sciences and
technologies, do we really have to move heaven and earth? Yes, exactly,
the Heavens and the Earth.

The hypothesis of this essay is that the word 'modern' designates two
sets of entirely different practices which must remain distinct if they are
to remain effective, but have recently begun to be confused. The first set
of practices, by 'translation', creates mixtures between entirely new types
of beings, hybrids of nature and culture. The second, by 'purification',
creates two entirely distinct ontological zones: that of human beings on

the one hand; that of <u>nonhumans</u> on the other. Without the first set, the practices of purification would be fruitless or pointless. Without the second, the work of translation would be slowed down, limited, or even ruled out. The first set corresponds to what I have called networks; the second to what I shall call the modern critical stance. The first, for example, would link in one continuous chain the chemistry of the upper atmosphere, scientific and industrial strategies, the preoccupations of heads of state, the anxieties of ecologists; the second would establish a partition between a natural world that has always been there, a society with predictable and stable interests and stakes, and a discourse that is independent of both reference and society.

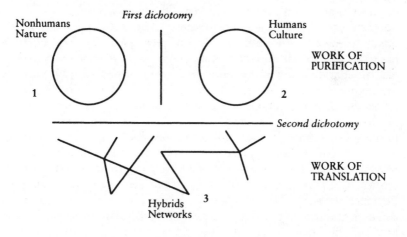

Figure 1.1 Purification and translation

So long as we consider these two practices of translation and purification separately, we are truly modern – that is, we willingly subscribe to the critical project, even though that project is developed only through the proliferation of hybrids down below. As soon as we direct our attention simultaneously to the work of purification and the work of hybridization, we immediately stop being wholly modern, and our future begins to change. At the same time we stop having been modern, because we become retrospectively aware that the two sets of practices have always already been at work in the historical period that is ending. Our past begins to change. Finally, if we have never been modern – at least in the way criticism tells the story – the tortuous relations that we have maintained with the other nature-cultures would also be transformed. Relativism, domination, imperialism, false consciousness, syncretism – all the problems that anthropologists summarize under the loose expression of 'Great Divide' – would be explained differently, thereby modifying comparative anthropology.

What link is there between the work of translation or mediation and that of purification? This is the question on which I should like to shed light. My hypothesis – which remains too crude – is that the second has made the first possible: the more we forbid ourselves to conceive of hybrids, the more possible their interbreeding becomes – such is the paradox of the moderns, which the exceptional situation in which we find ourselves today allows us finally to grasp. The second question has to do with premoderns, with the other types of culture. My hypothesis – once again too simple – is that by devoting themselves to conceiving of hybrids, the other cultures have excluded their proliferation. It is this disparity that would explain the Great Divide between Them – all the other cultures – and Us – the westerners – and would make it possible finally to solve the insoluble problem of relativism. The third question has to do with the current crisis: if modernity were so effective in its dual task of separation and proliferation, why would it weaken itself today by preventing us from being truly modern? Hence the final question, which is also the most difficult one: if we have stopped being modern, if we can no longer separate the work of proliferation from the work of purification, what are we going to become? Can we aspire to Enlightenment without modernity? My hypothesis – which, like the previous ones, is too coarse – is that we are going to have to slow down, reorient and regulate the proliferation of monsters by representing their existence officially. Will a different democracy become necessary? A democracy extended to things? To answer these questions, I shall have to sort out the premoderns, the moderns, and even the postmoderns in order to distinguish between their durable characteristics and their lethal ones.

Too many questions, as I am well aware, for an essay that has no excuse but its brevity. Nietzsche said that the big problems were like cold baths: you have to get out as fast as you got in.

2

□

CONSTITUTION

2.1 The Modern Constitution

Modernity is often defined in terms of humanism, either as a way of saluting the birth of 'man' or as a way of announcing his death. But this habit itself is modern, because it remains asymmetrical. It overlooks the simultaneous birth of 'nonhumanity' – things, or objects, or beasts – and the equally strange beginning of a crossed-out God, relegated to the sidelines. Modernity arises first from the conjoined creation of those three entities, and then from the masking of the conjoined birth and the separate treatment of the three communities while, underneath, hybrids continue to multiply as an effect of this separate treatment. The double separation is what we have to reconstruct: the separation between humans and nonhumans on the one hand, and between what happens 'above' and what happens 'below' on the other.

These separations could be compared to the division that distinguishes the judiciary from the executive branch of a government. This division is powerless to account for the multiple links, the intersecting influences, the continual negotiations between judges and politicians. Yet it would be a mistake to deny the effectiveness of the separation. The modern divide between the natural world and the social world has the same constitutional character, with one difference: up to now, no one has taken on the task of studying scientists and politicians in tandem, since no central vantage point has seemed to exist. In one sense, the fundamental articles of faith pertaining to the double separation have been so well drawn up that this separation has been viewed as a double ontological distinction. As soon as one outlines the symmetrical space and thereby reestablishes the common understanding that organizes the separation of natural and political powers, one ceases to be modern.

The common text that defines this understanding and this separation is called a constitution, as when we talk about amendments to the American constitution. Who is drafting such a text? For political constitutions, the task falls to jurists and Founding Fathers, but so far they have done only a third of the work, since they have left out both scientific power and the work of hybrids. For the nature of things, it is the scientists' task, but they have done only another third of the work, since they have pretended to forget about political power, and they have denied that hybrids have any role to play even as they multiply them. For the work of translation, writing the constitution is the task of those who study those strange networks that I have outlined above, but science students have fulfilled only half of their contract, since they do not explain the work of purification that is carried out above them and accounts for the proliferation of hybrids.

Who is to write the full constitution? As far as foreign collectives are concerned, anthropology has been pretty good at tackling everything at once. In fact, as we have seen, every ethnologist is capable of including within a single monograph the definition of the forces in play; the distribution of powers among human beings, gods, and nonhumans; the procedures for reaching agreements; the connections between religion and power; ancestors; cosmology; property rights; plant and animal taxonomies. The ethnologist will certainly not write three separate books: one dealing with knowledge, another with power, yet another with practices. She will write a single book, like the magnificent one in which Philippe Descola attempts to sum up the constitution of the Achuar of the Amazon region (Descola, [1986] 1993):

> Yet the Achuar have not completely subdued nature by the symbolic networks of domesticity. Granted, the cultural sphere is all-encompassing, since in it we find animals, plants and spirits which other Amerindian societies place in the realm of nature. The Achuar do not, therefore, share this antinomy between two closed and irremediably opposed worlds: the cultural world of human society and the natural world of animal society. And yet there is nevertheless a certain point at which the continuum of sociability breaks down, yielding to a wild world inexorably foreign to humans. Incomparably smaller than the realm of culture, this little piece of nature includes the set of things with which communication cannot be established. Opposite beings endowed with language [aents], of which humans are the most perfect incarnation, stand those things deprived of speech that inhabit parallel, inaccessible worlds. The inability to communicate is often ascribed to a lack of soul [wakan] that affects certain living species: most insects and fish, poultry, and numerous plants, which thus lead a mechanical, inconsequential existence. But the absence of communication is sometimes due to distance: the souls of stars and meteors,

infinitely far away and prodigiously mobile, remain deaf to human words.
[p. 399]

If an anthropology of the modern world were to exist its task would consist in describing in the same way how all the branches of our government are organized, including that of nature and the hard sciences, and in explaining how and why these branches diverge as well as accounting for the multiple arrangements that bring them together. The ethnologist of our world must take up her position at the common locus where roles, actions and abilities are distributed – those that make it possible to define one entity as animal or material and another as a free agent; one as endowed with consciousness, another as mechanical, and still another as unconscious and incompetent. Our ethnologist must even compare the always different ways of defining – or not defining – matter, law, consciousness and animals' souls, without using modern metaphysics as a vantage point. Just as the constitution of jurists defines the rights and duties of citizens and the State, the working of justice and the transfer of power, so this Constitution – which I shall spell with a capital C to distinguish it from the political ones – defines humans and nonhumans, their properties and their relations, their abilities and their groupings.

How can this Constitution be described? I have chosen to concentrate on an exemplary situation that arose at the very beginning of its drafting, in the middle of the seventeenth century, when the natural philosopher Robert Boyle and the political philosopher Thomas Hobbes were arguing over the distribution of scientific and political power. Such a choice might appear arbitrary if a remarkable book had not just come to grips with this double creation of a social context and a nature that escapes that very context. I shall use Boyle and Hobbes, along with their descendants and disciples, as a way of summarizing a much longer story – one that I cannot retrace here but one that others, better equipped than I, may want to pursue.

2.2 Boyle and His Objects

A book by Steven Shapin and Simon Schaffer (Shapin and Schaffer, 1985) marks the real beginning of a comparative anthropology that takes science seriously. At first glance, this book does nothing more than exemplify what has been the slogan of the Edinburgh school of science studies (Barnes and Shapin, 1979; Bloor, [1976] 1991) and of a great body of work in the social history of science (Shapin, 1982) and in the sociology of knowledge (Moscovici, 1977): 'questions of epistemology are also questions of social order'. It is impossible to do justice to either question if the two are separated, one assigned to departments of philosophy and the other to departments of sociology or political science. But Shapin and Schaffer push this general programme to the limit – first by displacing the historical beginning of this very divide between epistemology and sociology, and second, in part unwittingly, by ruining the privilege given to the social context in explaining the sciences.

> We have not referred to politics as something that happens solely outside of science and which can, so to speak, press in upon it. The experimental community [set up by Boyle] vigorously developed and deployed such boundary-speech, and we have sought to situate this speech historically and to explain why these conventionalized ways of talking developed. What we cannot do if we want to be serious about the historical nature of our inquiry is to use such actors' speech unthinkingly as an explanatory resource. The language that transports politics outside of science is precisely what we need to understand and explain. We find ourselves standing against much current sentiment in the history of science that holds that we should have less talk of the 'insides' and 'outsides' of science, that we have transcended such outmoded categories. Far from it; we have not yet begun to understand the issues involved. We still need to understand how such boundary-conventions developed: how, as a matter of historical record, scientific actors allocated items with respect to their boundaries (not ours), and how, as a matter of record, they behaved with respect to the items thus allocated. Nor should we take any one system of boundaries as belonging self-evidently to the thing that is called 'science.' (Shapin and Schaffer, 1985, p. 342)

In this long passage the authors do not show how the social context of England might justify the development of Boyle's physics and the failure of Hobbes's mathematical theories. They come to grips with the very basis of political philosophy. Far from 'situating Boyle's scientific works in their social context' or showing how politics 'presses in upon' scientific doctrines, they examine how Boyle and Hobbes fought to invent a science, a context, and a demarcation between the two. They are not prepared to explain the content by the context, since neither existed in this new way before Boyle and Hobbes reached their respective goals and settled their differences.

The beauty of Shapin and Schaffer's book stems from their success in unearthing Hobbes's scientific works – which had been neglected by political scientists, because they were embarrassed by the wild mathematical imaginings of their hero – and in rescuing from oblivion Boyle's political theories – which had been neglected by historians of science because they preferred to conceal their hero's organizational efforts. Instead of setting up an asymmetry, instead of distributing science to Boyle and political theory to Hobbes, Shapin and Schaffer outline a rather nice quadrant: Boyle has a science and a political theory; Hobbes has a political theory and a science. The quadrant would be uninteresting if the ideas of our two heroes were too far apart – if, for example, one were a philosopher after the fashion of Paracelsus and the other a Bodin-style lawmaker. But by good fortune, they agree on almost everything. They want a king, a Parliament, a docile and unified Church, and they are fervent subscribers to mechanistic philosophy. But even though both are thoroughgoing rationalists, their opinions diverge as to what can be expected from experimentation, from scientific reasoning, from political argument – and above all from the air pump, the real hero of the story. The disagreements between the two men, who agree on everything else, make them the ideal laboratory material, the perfect fruit flies for the new anthropology.

Boyle carefully refrained from talking about vacuum pumps. To put some order into the debates that followed the discovery of the Toricellian space at the top of a mercury tube inverted in a basin of the same substance, he claimed to be investigating only the weight of the air without taking sides in the dispute between plenists and vacuists. The apparatus he developed (modelled on Otto von Guericke's) that would permanently evacuate the air from a transparent glass container was, for the period – in terms of cost, complication and novelty – the equivalent of a major piece of equipment in contemporary physics. This was already Big Science. The great advantage of Boyle's installations was that they made it possible to see inside the glass walls and to introduce or even manipulate samples, owing to a series of ingeniously constructed lock chambers and covers. The pistons of the pump, the thick glass containers and the gaskets were not of adequate quality, so Boyle had to push technological research far enough, for instance, to be able to carry out the experiment he cared about most: that of the vacuum within a vacuum. He enclosed a Torricelli tube within the pump's glass enclosure and thus obtained an initial space at the top of the overturned tube. Then, by getting one of his technicians (who were invisible [Shapin, 1989]) to work the pump, he suppressed the weight of the air enough to bring down the level of the column, which descended nearly to the level of the mercury in the basin. Boyle undertook dozens of experiments within the confined chamber of his air pump, starting with attempts to detect the ether wind postulated by his adversaries, or to explain the cohesiveness of marble cylinders, or to suffocate small animals and put out candles – these experiments were later popularized by eighteenth-century parlour physics.

While a dozen civil wars were raging, Boyle chose a method of argument – that of opinion – that was held in contempt by the oldest scholastic tradition. Boyle and his colleagues abandoned the certainties of apodeictic reasoning in favour of a doxa. This doxa was not the raving imagination of the credulous masses, but a new mechanism for winning the support of one's peers. Instead of seeking to ground his work in logic, mathematics or rhetoric, Boyle relied on a parajuridical metaphor: credible, trustworthy, well-to-do witnesses gathered at the scene of the action can attest to the existence of a fact, the matter of fact, even if they do not know its true nature. So he invented the empirical style that we still use today (Shapin, 1984).

Boyle did not seek these gentlemen's opinion, but rather their observation of a phenomenon produced artificially in the closed and protected space of a laboratory (Shapin, 1990). Ironically, the key question of the constructivists – are facts thoroughly constructed in the laboratory? (Woolgar, 1988) – is precisely the question that Boyle raised and resolved. Yes, the facts are indeed constructed in the new installation of the laboratory and through the artificial intermediary of the air pump. The level does descend in the Torricelli tube that has been inserted into the transparent enclosure of a pump operated by breathless technicians. '*Les faits sont faits*': 'Facts are fabricated,' as Gaston Bachelard would say. But are facts that have been constructed by man artifactual for that reason? No: for Boyle, just like Hobbes, extends God's 'constructivism' to man. God knows things because He creates them (Funkenstein, 1986). We know the nature of the facts because we have developed them in circumstances that are under our complete control. Our weakness becomes a strength, provided that we limit knowledge to the instrumentalized nature of the facts and leave aside the interpretation of causes. Once again, Boyle turns a flaw – we produce only matters of fact that are created in laboratories and have only local value – into a decisive advantage: these facts will never be modified, whatever may happen elsewhere in theory, metaphysics, religion, politics or logic.

2.3 Hobbes and His Subjects

Hobbes rejected Boyle's entire theatre of proof. Like Boyle, Hobbes too wanted to bring an end to the civil war; he too wanted to abandon free interpretation of the Bible on the part of clerics and the people alike. But he meant to reach his goal by a unification of the Body Politic. The Sovereign created by the contract, 'that *Mortall God*, to which we owe, under the *Immortal God*, our peace and defence' (Hobbes, [1651] 1947, p. 89), is only the representative of the multitude. 'For it is the *Unity* of the Represener, not the *Unity* of the Represented, that maketh the Person *One*' (p. 85). Hobbes was obsessed by the unity of the Person who is, as he puts it, the Actor of which we citizens are the Authors. It is because of this unity that there can be no transcendence. Civil wars will rage as long as there exist supernatural entities that citizens feel they have a right to petition when they are persecuted by the authorities of this lower world. The loyalty of the old medieval society – to God and King – is no longer possible if all people can petition God directly, or designate their own King. Hobbes wanted to wipe the slate clean of all appeals to entities higher than civil authority. He wanted to rediscover Catholic unity while at the same time closing off any access to divine transcendence.

For Hobbes, Power is Knowledge, which amounts to saying that there can exist only one Knowledge and only one Power if civil wars are to be brought to an end. This is why the major portion of *Leviathan* is devoted to an exegesis of the Old and New Testaments. One of the great dangers for civil peace comes from the belief in immaterial bodies such as spirits, phantoms or souls, to which people appeal against the judgements of civil power. Antigone might be dangerous when she proclaims the superiority of piety over Creon's 'reasons of State'; the egalitarians, the Levellers and the Diggers are much more so when they invoke the active powers of matter and the free interpretation of the Bible in order to disobey their legitimate princes. Inert and mechanical matter is as essential to civil peace as a purely symbolic interpretation of the Bible. In both cases, it behoves us to avoid at all costs the possibility that the factions may invoke a higher Entity – Nature or God – which the Sovereign does not fully control.

This reductionism does not lead to a totalitarian State, since Hobbes applies it to the Republic itself: the Sovereign is never anything but an Actor designated by the social contract. There is no divine law or higher agency that the Sovereign might invoke in order to act as he wishes and dismantle the Leviathan. In this new regime in which Knowledge equals Power, everything is cut down to size: the Sovereign, God, matter, and the multitude. Hobbes even rules out turning his own science of the State into an invocation of transcendence. He arrives at all his scientific results not by opinion, observation or revelation but by a mathematical demonstration, the only method of argument capable of compelling everyone's assent; and he accomplishes this demonstration not by making transcendental calculations, like Plato's King, but by using a purely computational instrument, the Mechanical Brain, a computer before its time. Even the famous social contract is only the sum of a calculation reached abruptly and simultaneously by all the terrorized citizens who are seeking to liberate themselves from the state of nature. Such is Hobbes's generalized constructivism designed to end civil war: no transcendence whatsoever, no recourse to God, or to active matter, or to Power by Divine Right, or even to mathematical Ideas.

All the elements are now in place for the confrontation between Hobbes and Boyle. After Hobbes has reduced and reunified the Body Politic, along comes the Royal Society to divide everything up again: some gentlemen proclaim the right to have an independent opinion, in a closed space, the laboratory, over which the State has no control. And when these troublemakers find themselves in agreement, it is not on the basis of a mathematical demonstration that everyone would be compelled to accept, but on the basis of experiments observed by the deceptive senses, experiments that remain inexplicable and inconclusive. Worse still, this new coterie chooses to concentrate its work on an air pump that once again produces immaterial bodies, the vacuum – as if Hobbes had not had enough trouble getting rid of phantoms and spirits! And here we are again, Hobbes worries, right in the middle of a civil war! We are no longer to be subjected to the Levellers and the Diggers, who challenged the King's authority in the name of their personal interpretation of God and of the properties of matter (they have been properly exterminated), but we are going to have to put up with this new clique of scholars who are going to start challenging everyone's authority in the name of Nature by invoking wholly fabricated laboratory events! If you allow experiments to produce their own matters of fact, and if these allow the vacuum to be infiltrated into the air pump and, from there, into natural philosophy, then you will divide authority again: the immaterial spirits will incite everyone to revolt by offering a court of appeal for frustrations. Knowledge and Power will be separated once more. You will 'see double', as Hobbes put it. Such are the warnings he addresses to the King in denouncing the goings-on of the Royal Society.

2.4 The Mediation of the Laboratory

This political interpretation of Hobbes's plenism does not suffice to make Shapin and Schaffer's book a solid foundation for comparative anthropology. Any good historian of ideas could have done the same job. But in three decisive chapters our authors leave the confines of intellectual history and pass from the world of opinions and argument to the world of practices and networks. For the first time in science studies, all ideas pertaining to God, the King, Matter, Miracles and Morality are translated, transcribed, and forced to pass through the practice of making an instrument work. Before Shapin and Schaffer, other historians of science had studied scientific practice; other historians had studied the religious, political and cultural context of science. No one, before Shapin and Schaffer, had been capable of doing both at once.

Just as Boyle succeeds in transforming his tinkering about with a jerry-built air pump into the partial assent of gentlemen with respect to facts that have become indisputable, so Shapin and Schaffer manage to explain how and why discussions dealing with the Body Politic, God and His miracles, matter and its power, have to be translated through the air pump. This mystery has never been cleared up by those seeking a contextualist explanation for the sciences. Contextualists start from the principle that a social macro-context exists – England, the dynastic quarrel, Capitalism, Revolution, Merchants, the Church – and that this context in some way influences, forms, reflects, has repercussions for, and exercises pressure on 'ideas about' matter, the air's spring, vacuums, and Torricelli tubes. But they never explain the prior establishment of a link connecting God, the King, Parliament, and some bird suffocating in the transparent closed chamber of a pump whose air is being removed by means of a crank operated by a technician. How can the bird's experience translate, displace, transport, distort all the other controversies, in such a way that those who master the pump also master the King, God, and the entire context?

Hobbes indeed seeks to get round everything that has to do with experimental work, but Boyle forces the discussion to proceed by way of a set of sordid details involving the leaks, gaskets and cranks of his machine. In the same way, philosophers of science and historians of ideas would like to avoid the world of the laboratory, that repugnant kitchen in which concepts are smothered with trivia (Cunningham and Williams, 1992; Knorr, 1981; Latour and Woolgar, [1979] 1986; Pickering, 1992; Traweek, 1988). Shapin and Schaffer force their analyses to hinge on the object, on a certain leak, a particular gasket in the air pump. The practice of fabricating objects is restored to the dominant place it had lost with the modern critical stance. Their book is not empirical simply because of its abundant details; it is empirical because it undertakes the archaeology of that new object that is born in the seventeenth century in the laboratory. Shapin and Schaffer, like Ian Hacking (Hacking, 1983), do in a quasi-ethnographic way what philosophers of science now do scarcely at all: they show the realistic foundations of the sciences. But rather than speaking of the external reality 'out there', they anchor the indisputable reality of science 'down there', on the bench.

The experiments don't go very well. The pump leaks. It has to be patched up. Those who are incapable of explaining the irruption of objects into the human collective, along with all the manipulations and practices that objects require, are not anthropologists, for what has constituted the most fundamental aspect of our culture, since Boyle's day, eludes them: we live in communities whose social bond comes from objects fabricated in laboratories; ideas have been replaced by practices, apodeictic reasoning by a controlled doxa, and universal agreement by groups of colleagues. The lovely order that Hobbes was trying to recover is annihilated by the multiplication of private spaces where the transcendental origin of facts is proclaimed – facts that have been fabricated by man yet are no one's handiwork, facts that have no causality yet can be explained.

How can a society be made to hold together peacefully, Hobbes asks indignantly, on the pathetic foundation of matters of fact? He is particularly annoyed by the relative change in the scale of phenomena. According to Boyle, the big questions concerning matter and divine power can be subjected to experimental resolution, and this resolution will be partial and modest. Now Hobbes rejects the possibility of the vacuum for ontological and political reasons of primary philosophy, and he continues to allege the existence of an invisible ether that must be present, even when Boyle's worker is too out of breath to operate his pump. In other words, he demands a macroscopic response to his 'macro-'arguments, a demonstration that would prove that his ontology is not necessary, that the vacuum is politically acceptable. Now what does Boyle do in response? He chooses, on the contrary, to make his experiment more sophisticated, to show the effect on a detector – a mere chicken feather! – of the ether wind postulated by Hobbes in the hope of invalidating his detractor's theory (Shapin and Schaffer, 1985, p. 182). Ridiculous! Hobbes raises a fundamental problem of political philosophy, and his theories are to be refuted by a feather in a glass chamber inside Boyle's mansion! Of course, the feather doesn't move at all, and Boyle draws the conclusion that Hobbes is wrong, that there is no ether wind. However, Hobbes cannot be wrong, because he refuses to admit that the phenomenon he is talking about can be produced on a scale other than that of the Republic as a whole. He denies what is to become the essential characteristic of modern power: the change in scale and the displacements that are presupposed by laboratory work (Latour, 1983). Boyle, a new Puss in Boots, now has only to pounce on the Ogre, who has just been reduced to the size of a mouse.

2.5 The Testimony of Nonhumans

Boyle's innovation is striking. Against Hobbes's judgement, he takes possession of the old repertoire of penal law and biblical exegesis, but he does so in order to apply them to the testimony of the things put to the test in the laboratory. As Shapin and Schaffer write:

> Sprat and Boyle appealed to 'the practice of our courts of justice here in England' to sustain the moral certainty of their conclusions and to support the argument that the multiplication of witnesses allowed 'a concurrence of

such probabilities.' Boyle used the provision of Clarendon's 1661 Treason
Act, in which, he said, two witnesses were necessary to convict. So the legal
and priestly models of authority through witnessing were fundamental
resources for the experimenters. Reliable witnesses were ipso facto the
members of a trustworthy community: Papists, atheists, and sectaries
found their stories challenged, the social status of a witness sustained his
credibility, and the concurring voices of many witnesses put the extremists
to flight. Hobbes challenged the basis of this practice: once again, he
displayed the form of life that sustained witnessing as an ineffective and
subversive enterprise. (Shapin and Schaffer, 1985, p. 327)

At first glance, Boyle's repertoire does not contribute much that is new.
Scholars, monks, jurists and scribes had been developing all those
resources for a millennium and more. What is new, however, is their
point of application. Earlier, the witnesses had always been human or
divine – never nonhuman. The texts had been written by men or inspired
by God – never inspired or written by nonhumans. The law courts had
seen countless human and divine trials come and go – never affairs that
called into question the behaviour of nonhumans in a laboratory
transformed into a court of justice. Yet for Boyle, laboratory experiments
carry more authority than unconfirmed depositions by honourable
witnesses:

> 'The pressure of the water in our recited experiment [on the diver's bell]
> having manifest effects upon inanimate bodies, which are not capable of
> prepossessions, or giving us partial informations, will have much more
> weight with unprejudiced persons, than the suspicious, and sometimes
> disagreeing accounts of ignorant divers, whom prejudicate opinions may
> much sway, and whose very sensations, as those of other vulgar men, may
> be influenced by predispositions, and so many other circumstances, that
> they may easily give occasion to mistakes.' [Shapin and Schaffer, 1985,
> p. 218]

Here in Boyle's text we witness the intervention of a new actor
recognized by the new Constitution: inert bodies, incapable of will and
bias but capable of showing, signing, writing, and scribbling on
laboratory instruments before trustworthy witnesses. These nonhumans,
lacking souls but endowed with meaning, are even more reliable than
ordinary mortals, to whom will is attributed but who lack the capacity to
indicate phenomena in a reliable way. According to the Constitution, in
case of doubt, humans are better off appealing to nonhumans. Endowed
with their new semiotic powers, the latter contribute to a new form of
text, the experimental science article, a hybrid between the age-old style
of biblical exegesis – which has previously been applied only to the

Scriptures and classical texts – and the new instrument that produces
new inscriptions. From this point on, witnesses will pursue their
discussions around the air pump in its enclosed space, discussions about
the meaningful behaviour of nonhumans. The old hermeneutics will
persist, but it will add to its parchments the shaky signature of scientific
instruments (Latour and De Noblet, 1985; Law and Fyfe, 1988; Lynch
and Woolgar, 1990). With a law court thus renewed, all the other powers
will be overthrown, and this is what makes Hobbes so upset; however,
the overturning is possible only if all connections with the political and
religious branches of government become impossible.

Shapin and Schaffer pursue their discussion of objects, laboratories,
capacities, and changes of scale to its extreme consequences. If science is
based not on ideas but on a practice, if it is located not outside but inside
the transparent chamber of the air pump, and if it takes place within the
private space of the experimental community, then how does it reach
'everywhere'? How does it become as universal as 'Boyle's laws' or
'Newton's laws'? The answer is that it never become universal – not, at
least, in the epistemologists' terms! Its network is extended and
stabilized. This expansion is brilliantly demonstrated in a chapter which,
like the work of Harry Collins (Collins, 1985) or Trevor Pinch (Pinch,
1986) offers a striking example of the fruitfulness of the new science
studies. By following the reproduction of each prototype air pump
throughout Europe, and the progressive transformation of a piece of
costly, not very reliable and quite cumbersome equipment, into a cheap
black box that gradually becomes standard equipment in every laboratory, the authors bring the universal application of a law of physics back
within a network of standardized practices. Unquestionably, Boyle's
interpretation of the air's spring is propagated – but its speed of
propagation is exactly equivalent to the rate at which the community of
experimenters and their equipment develop. No science can exit from the
network of its practice. The weight of air is indeed always a universal,
but a universal in a network. Owing to the extension of this network,
competences and equipment can become sufficiently routine for production of the vacuum to become as invisible as the air we breathe; but
universal in the old sense? Never.

2.6 The Double Artifact of the Laboratory and the Leviathan

How far does the symmetry hold between Hobbes's invention and
Boyle's? Shapin and Schaffer are not clear on this point. At first sight,
however, it seems that Hobbes and his disciples created the chief
resources that are available to us for speaking about power ('representa-

tion', 'sovereign', 'contract', 'property', 'citizens'), while Boyle and his successors developed one of the major repertoires for speaking about nature ('experiment', 'fact', 'evidence', 'colleagues'). It should thus seem also clear that we are dealing not with two separate inventions but with only one, a division of power between the two protagonists, to Hobbes, the politics and to Boyle, the sciences. This, however, is not the conclusion drawn by Shapin and Schaffer. After having had the stroke of genius that led them to compare the experimental practice and political organization of two major figures from the very beginning of the modern era, they back off and hesitate to treat Hobbes and his politics in the same way as they had treated Boyle and his science. Strangely enough, they seem to adhere more steadfastly to the political repertoire than to the scientific one.

Yet Shapin and Schaffer unintentionally displace the traditional centre of reference of the modern critique downward. If science is based on forms of life, practices, laboratories and networks, then where is it to be situated? Certainly not on the side of things-in-themselves, since the facts are fabricated. But it cannot be situated, either, on the side of the subject – or whatever name one wants to give this side: society, brain, spirit, language game, epistemes or culture. The suffocating bird, the marble cylinders, the descending mercury are not our own creations, they are not made out of thin air, not of social relations, not of human categories. Must we then place the practice of science right in the middle of the line that connects the Object Pole to the Subject Pole? Is this practice a hybrid, or a mixture of the two? Part object and part subject? Or is it necessary to invent a new position for this strange generation of both a political context and a scientific content?

The authors do not give us a definitive answer to these questions as if they had failed to do justice to their own discovery. Just as Hobbes and Boyle agree on everything except how to carry out experiments, the authors, who agree on everything, disagree on how to deal with the 'social' context – that is, Hobbes's symmetrical invention of a human capable of being represented. The last chapters of the book waver between a Hobbesian explanation of the authors' own work and a Boylian point of view. This tension only makes their work more interesting, and it supplies the anthropology of science with a new line of ideally suited fruit flies, since they differ by only a few traits. Shapin and Schaffer consider Hobbes's macro-social explanations relative to Boyle's science more convincing than Boyle's arguments refuting Hobbes! Trained in the framework of the social study of sciences, they seem to accept the limitations imposed by the Edinburgh school: if all questions of epistemology are questions of social order, this is because, when all is said and done, the social context contains as one of its subsets the definition of what counts as good science. Such an asymmetry renders Shapin and Schaffer less well equipped to deconstruct the macro-social context than Nature 'out there'. They seem to believe that a society 'up there' actually exists, and that it accounts for the failure of Hobbes's programme. Or – more precisely – they do not manage to settle the question, cancelling out in their conclusion what they had demonstrated in Chapter 7, and cancelling out their own argument yet again in the very last sentence of the book:

> Neither our scientific knowledge, nor the constitution of our society, nor traditional statements about the connections between our society and our knowledge are taken for granted any longer. As we come to recognize the conventional and artifactual status of our forms of knowing, we put ourselves in a position to realize that it is ourselves and not reality that is responsible for what we know. Knowledge, as much as the State, is the product of human actions. Hobbes was right. [p. 344]

No, Hobbes was wrong. How could he have been right, when he was the one who invented the monist society in which Knowledge and Power are one and the same thing? How can such a crude theory be used to explain Boyle's invention of an absolute dichotomy between the production of knowledge of facts and politics? Yes, 'knowledge, as much as the State, is the product of human actions', but that is precisely why Boyle's political invention is much more refined than Hobbes's sociology of science. If we are to understand the final obstacle separating us from an anthropology of science, we have to deconstruct Hobbes's constitutional invention according to which there is such a thing as a macro-society much sturdier and more robust than Nature.

Hobbes invents the naked calculating citizen, whose rights are limited to possessing and to being represented by the artificial construction of the Sovereign. He also creates the language according to which Power equals Knowledge, an equation that is at the root of the entire modern Realpolitik. Furthermore, he offers a set of terms for analyzing human interests which, along with Machiavelli's, remains the basic vocabulary for all of sociology today. In other words, even though Shapin and Schaffer take great care to use the expression 'scientific fact' not as a resource but rather as a historical and political invention, they take no such precautions where political language itself is concerned. They use the words 'power', 'interest' and 'politics' in all innocence (Chapter 7). Yet who invented these words, with their modern meaning? Hobbes! Our authors are thus 'seeing double' themselves, and walking sideways, criticizing science but swallowing politics as the only valid source of explanation. Now who offers us this asymmetric way of explaining knowledge through power? Hobbes again, with his construction of a monist macro-structure in which knowledge has a place only in support of the social order. The authors offer a masterful deconstruction of the evolution, diffusion and popularization of the air pump. Why, then, do they not deconstruct the evolution, diffusion and popularization of 'power' or 'force'? Is 'force' less problematic than the air's spring? If nature and epistemology are not made up of transhistoric entities, then neither are history and sociology – unless one adopts some authors' asymmetrical posture and agrees to be simultaneously constructivist where nature is concerned and realist where society is concerned (Collins and Yearley, 1992)! But it is not very probable that the air's spring has a more political basis than English society itself . . .

2.7 Scientific Representation and Political Representation

If, unlike Shapin and Schaffer themselves, we pursue the logic of their book to the end, we understand the symmetry of the work achieved simultaneously by Hobbes and Boyle, and we might locate the practice of science that they have described. Boyle is not simply creating a scientific discourse while Hobbes is doing the same thing for politics; Boyle is creating a political discourse from which politics is to be excluded, while Hobbes is imagining a scientific politics from which experimental science has to be excluded. In other words, they are inventing our modern world, a world in which the representation of things through the intermediary of the laboratory is forever dissociated from the representation of citizens through the intermediary of the social contract. So it is not at all by oversight that political philosophers have ignored Hobbes's science, while historians of science have ignored Boyle's positions on the politics of science. All of them had to 'see double' from Hobbes's and Boyle's day on, and not establish direct relations between the representation of nonhumans and the representation of humans, between the artificiality of facts and the artificiality of the Body Politic. The word 'representation' is the same, but the controversy between Hobbes and Boyle renders any likeness between the two senses of the word unthinkable. Today, now that we are no longer entirely modern, these two senses are moving closer together again.

The link between epistemology and social order now takes a completely new meaning. The two branches of government that Boyle and Hobbes develop, each on his own side, possess authority only if they are clearly separated: Hobbes's State is impotent without science and technology, but Hobbes speaks only of the representation of naked citizens; Boyle's science is impotent without a precise delimitation of the religious, political and scientific spheres, and that is why he makes such an effort to counteract Hobbes's monism. They are like a pair of Founding Fathers, acting in concert to promote one and the same innovation in political theory: the representation of nonhumans belongs to science, but science is not allowed to appeal to politics; the representation of citizens belongs to politics, but politics is not allowed to have any relation to the nonhumans produced and mobilized by science and technology. Hobbes and Boyle quarrel in order to define the two resources that we continue to use unthinkingly, and the intensity of their double battle is highly indicative of the novelty of what they are inventing.

Hobbes defines a naked and calculating citizen who constitutes the Leviathan, a mortal god, an artificial creature. On what does the Leviathan depend? On the calculation of human atoms that leads to the contract that decides on the irreversible composition of the strength of all in the hands of a single one. In what does this strength consist? In the authorization granted by all naked citizens to a single one to speak in their name. Who is acting when that one acts? We are, we who have definitively delegated our power to him. The Republic is a paradoxical artificial creature composed of citizens united only by the authorization given to one of them to represent them all. Does the Sovereign speak in his own name, or in the name of those who empower him? This is an insoluble question with which modern political philosophy will grapple endlessly. It is indeed the Sovereign who speaks, but it is the citizens who are speaking through him. He becomes their spokesperson, their persona, their personification. He translates them; therefore he may betray them. They empower him: therefore they may impeach him. The Leviathan is made up only of citizens, calculations, agreements or disputes. In short, it is made up of nothing but social relations. Or rather, thanks to Hobbes and his successors, we are beginning to understand what is meant by social relations, powers, forces, societies.

But Boyle defines an even stranger artifact. He invents the laboratory within which artificial machines create phenomena out of whole cloth. Even though they are artificial, costly and hard to reproduce, and despite the small number of trained and reliable witnesses, these facts indeed represent nature as it is. The facts are produced and represented in the laboratory, in scientific writings; they are recognized and vouched for by the nascent community of witnesses. Scientists are scrupulous representatives of the facts. Who is speaking when they speak? The facts themselves, beyond all question, but also their authorized spokespersons. Who is speaking, then, nature or human beings? This is another insoluble question with which the modern philosophy of science will wrestle over the course of three centuries. In themselves, facts are mute; natural forces are brute mechanisms. Yet the scientists declare that they themselves are not speaking; rather, facts speak for themselves. These mute entities are thus capable of speaking, writing, signifying within the artificial chamber of the laboratory or inside the even more rarefied chamber of the vacuum pump. Little groups of gentlemen take testimony from natural forces, and they testify to each other that they are not betraying but translating the silent behaviour of objects. With Boyle and his successors, we begin to conceive of what a natural force is, an object that is mute but endowed or entrusted with meaning.

In their common debate, Hobbes's and Boyle's descendants offer us the resources we have used up to now: on the one hand, social force and power; on the other, natural force and mechanism. On the one hand, the subject of law; on the other, the object of science. The political spokespersons come to represent the quarrelsome and calculating multitude of citizens; the scientific spokespersons come to represent the mute and material multitude of objects. The former translate their principals, who cannot all speak at once; the latter translate their constituents, who are mute from birth. The former can betray; so can the latter. In the seventeenth century, the symmetry is still visible; the two camps are still arguing through spokespersons, each accusing the other of multiplying the sources of conflict. Only a little effort is now required for their common origin to become invisible, for there to be no more spokesperson except on the side of human beings, and for the scientists' mediation to become invisible. Soon the word 'representation' will take on two different meanings, according to whether elected agents or things are at stake. Epistemology and political science will go their opposite ways.

2.8 The Constitutional Guarantees of the Moderns

If the modern Constitution invents a separation between the scientific power charged with representing things and the political power charged with representing subjects, let us not draw the conclusion that from now on subjects are far removed from things. On the contrary. In his *Leviathan*, Hobbes simultaneously redraws physics, theology, psychology, law, biblical exegesis and political science. In his writing and his correspondence, Boyle simultaneously redesigns scientific rhetoric, theology, scientific politics, and the hermeneutics of facts. Together, they describe how God must rule, how the new King of England must legislate, how the spirits or the angels should act, what the properties of matter are, how nature is to be interrogated, what the boundaries of scientific or political discussion must be, how to keep the lower orders on a tight rein, what the rights and duties of women are, what is to be expected of mathematics. In practice, then, they are situated within the old anthropological matrix; they divide up the capacities of things and people, and they do not yet establish any separation between a pure social force and a pure natural mechanism.

Here lies the entire modern paradox. If we consider hybrids, we are dealing only with mixtures of nature and culture; if we consider the work of purification, we confront a total separation between nature and culture. It is the relation between these two tasks that I am seeking to understand. While both Boyle and Hobbes are meddling in politics and religion and technology and morality and science and law, they are also dividing up the tasks to the extent that the one restricts himself to the science of things and the other to the politics of men. What is the intimate relation between their two movements? Is purification necessary to allow for proliferation? Must there be hundreds of hybrids in order for a simply human politics and simply natural things to exist? Is an absolute distinction required between the two movements in order for both to remain effective? How can the power of this arrangement be explained? What, then, is the secret of the modern world? In an attempt to grasp the answers, we have to generalize the results achieved by Shapin and Schaffer and define the complete Constitution, of which Hobbes and Boyle wrote only one of the early drafts. To do so I have none of the historical skills of my colleagues and I will have to rely on what is, of necessity, a speculative exercise imagining that such a Constitution has indeed been drafted by conscious agents trying to build from scratch a functional system of checks and balances.

As with any Constitution, this one has to be measured by the guarantees it offers. The natural power that Boyle and his many scientific descendants defined in opposition to Hobbes, the power that allows mute objects to speak through the intermediary of loyal and disciplined scientific spokespersons, offers a significant guarantee: it is not men who make Nature; Nature has always existed and has always already been there; we are only discovering its secrets. The political power that Hobbes and his many political descendants define in opposition to Boyle has citizens speak with one voice through the translation and betrayal of a sovereign, who says only what they say. This power offers an equally significant guarantee: human beings, and only human beings, are the ones who construct society and freely determine their own destiny.

If, after the fashion of modern political philosophy, we consider these two guarantees separately, they remain incomprehensible. If Nature is not made by or for human beings, then it remains foreign, forever remote and hostile. Nature's very transcendence overwhelms us, or renders it inaccessible. Symmetrically, if society is made only by and for humans, the Leviathan, an artificial creature of which we are at once the form and the matter, cannot stand up. Its very immanence destroys it at once in the war of every man against every man. But these two constitutional guarantees must not be taken separately, as if the first assured the nonhumanity of Nature and the second the humanity of the social sphere. They were created together. They reinforce each other. The first and second guarantees serve as counterweight to one another, as checks and balances. They are nothing but the two branches of a single new government.

If we now consider them together, not separately, we note that the guarantees are reversed. Boyle and his descendants are not simply saying that the Laws of Nature escape our grasp; they are also fabricating these laws in the laboratory. Despite their artificial construction inside the vacuum pump (such is the phase of mediation or translation), the facts completely escape all human fabrication (such is the phase of purification). Hobbes and his descendants are not declaring simply that men make their own society by sheer force, but that the Leviathan is durable and solid, massive and powerful; that it mobilizes commerce, inventions, and the arts; and that the Sovereign holds the well-tempered steel sword and the golden sceptre in his hand. Despite its human construction, the Leviathan infinitely surpasses the humans who created it, for in its pores, its vessels, its tissues, it mobilizes the countless goods and objects that give it consistency and durability. Yet despite the solidity procured by the mobilization of things (as revealed by the work of mediation), we alone are the ones who constitute it freely by the sheer force of our reasoning – we poor, naked, unarmed citizens (as demonstrated by the work of purification).

But these two guarantees are contradictory, not only mutually but internally, since each plays simultaneously on transcendence and immanence. Boyle and his countless successors go on and on both constructing Nature artificially and stating that they are discovering it; Hobbes and the newly defined citizens go on and on constructing the Leviathan by dint of calculation and social force, but they recruit more and more objects in order to make it last. Are they lying? Deceiving themselves? Deceiving us? No, for they add a third constitutional guarantee: there shall exist a complete separation between the natural world (constructed, nevertheless, by man) and the social world (sustained, nevertheless, by things); secondly, there shall exist a total separation between the work of hybrids and the work of purification. The first two guarantees are contradictory only as long as the third does not keep them apart for ever, as long as it does not turn an overly patent symmetry into two contradictory asymmetries that practice resolves but can never express.

FIRST PARADOX

Nature is not our construction;   Society is our free construction;
it is transcendent and   it is immanent to our action.
surpasses us infinitely.

SECOND PARADOX

Nature is our artificial   Society is not our construction;
construction in the laboratory;   it is transcendent and surpasses
it is immanent.   us infinitely.

CONSTITUTION

First guarantee: even though we   Second guarantee: even though we
construct Nature, Nature is as if   do not construct Society, Society
we did not construct it.   is as if we did construct it.

Third guarantee: Nature and Society
must remain absolutely distinct: the
work of purification must remain absolutely
distinct from the work of mediation.

Figure 2.1 The paradoxes of Nature and Society

It will take many more authors, many more institutions, many more rules, to complete the movement sketched out by the exemplary dispute between Hobbes and Boyle. But the overall structure is now easy to grasp: the three guarantees taken together will allow the moderns a change in scale. They are going to be able to make Nature intervene at every point in the fabrication of their societies while they go right on attributing to Nature its radical transcendence; they are going to be able to become the only actors in their own political destiny, while they go right on making their society hold together by mobilizing Nature. On the one hand, the transcendence of Nature will not prevent its social immanence; on the other, the immanence of the social will not prevent the Leviathan from remaining transcendent. We must admit that this is a rather neat construction that makes it possible to do everything without being limited by anything. It is not surprising that this Constitution should have made it possible, as people used to say, to 'liberate productive forces. . .'

2.9 The Fourth Guarantee: The Crossed-out God

It was necessary, however, to avoid seeing an overly perfect symmetry between the two guarantees of the Constitution, which would have prevented that duo from giving its all. A fourth guarantee had to settle the question of God by removing Him for ever from the dual social and natural construction, while leaving Him presentable and usable nevertheless. Hobbes's and Boyle's followers succeeded in carrying out this task – the former by ridding Nature of any divine presence, the latter by ridding Society of any divine origin. Scientific power 'no longer needed this hypothesis'; as for statesmen, they could fabricate the 'mortal god' of the Leviathan without troubling themselves further about the immortal God whose Scripture was now interpreted only figuratively by the Sovereign. No one is truly modern who does not agree to keep God from interfering with Natural Law as well as with the laws of the Republic. God becomes the crossed-out God of metaphysics, as different from the premodern God of the Christians as the Nature constructed in the laboratory is from the ancient *phusis* or the Society invented by sociologists from the old anthropological collective and its crowds of nonhumans.

But an overly thorough distancing would have deprived the moderns of a critical resource they needed to complete their mechanism. The Nature-and-Society twins would have been left hanging in the void, and no one would have been able to decide, in case of conflict between the two branches of government, which one should win out over the other. Worse still, their symmetry would have been excessively obvious. If I am allowed to go on with the convenient fiction that this Constitution is drafted by some conscious agent endowed with will, foresight and cunning I could say that everything happens as if the moderns had applied the same doubling to the crossed-out God that they had used on Nature and Society. His transcendence distanced Him infinitely, so that He disturbed neither the free play of nature nor that of society, but the right was nevertheless reserved to appeal to that transcendence in case of conflict between the laws of Nature and those of Society. Modern men and women could thus be atheists even while remaining religious. They could invade the material world and freely re-create the social world, but without experiencing the feeling of an orphaned demiurge abandoned by all.

Reinterpretation of the ancient Christian theological themes made it possible to bring God's transcendence and His immanence into play simultaneously. But this lengthy task of the sixteenth-century Reformation would have produced very different results had it not got mixed up with the task of the seventeenth century, the conjoined invention of scientific facts and citizens (Eisenstein, 1979). Spirituality was reinvented: the all-powerful God could descend into men's heart of hearts without intervening in any way in their external affairs. A wholly individual and wholly spiritual religion made it possible to criticize both the ascendancy of science and that of society, without needing to bring God into either. The moderns could now be both secular and pious at the same time (Weber, [1920] 1958). This last constitutional guarantee was given not by a supreme God but by an absent God – yet His absence did not prevent people from calling on Him at will in the privacy of their own hearts. His position became literally ideal, since He was bracketed twice over, once in metaphysics and again in spirituality. He would no longer interfere in any way with the development of the moderns, but He remained effective and helpful within the spirit of humans alone.

A threefold transcendence and a threefold immanence in a crisscrossed schema that locks in all the possibilities: this is where I locate the power of the moderns. They have not made Nature; they make Society; they make Nature; they have not made Society; they have not made either, God has made everything; God has made nothing, they have made everything. There is no way we can understand the moderns if we do not see that the four guarantees serve as checks and balances for one another. The first two make it possible to alternate the sources of power by moving directly from pure natural force to pure political force, and vice versa. The third guarantee rules out any contamination between what belongs to Nature and what belongs to politics, even though the first two guarantees allow a rapid alternation between the two. Might the contradiction between the third, which separates, and the first two, which alternate, be too obvious? No, because the fourth constitutional guarantee establishes as arbiter an infinitely remote God who is simultaneously totally impotent and the sovereign judge.

If I am right in this outline of the Constitution, modernity has nothing to do with the invention of humanism, with the emergence of the sciences, with the secularization of society, or with the mechanization of the world. Its originality and its strength come from the conjoined production of these three pairings of transcendence and immanence, across a long history of which I have presented only one stage via the figures of Hobbes and Boyle. The essential point of this modern Constitution is that it renders the work of mediation that assembles hybrids invisible, unthinkable, unrepresentable. Does this lack of representation limit the work of mediation in any way? No, for the modern world would immediately cease to function. Like all other collectives it lives on that blending. On the contrary (and here the beauty of the mechanism comes to light), *the modern Constitution allows the expanded proliferation of the hybrids whose existence, whose very possibility, it denies.* By playing three times in a row on the same alternation between transcendence and immanence, the moderns can mobilize Nature, objectify the social, and feel the spiritual presence of God, even while firmly maintaining that Nature escapes us, that Society is our own work, and that God no longer intervenes. Who could have resisted such a construction? Truly exceptional events must have weakened this powerful mechanism for me to be able to describe it today with an ethnologist's detachment for a world that is in the process of disappearing.

2.10 The Power of the Modern Critique

At the very moment when the moderns' critical capacities are waning, it is useful to take the measure, one last time, of their prodigious efficacity.

Freed from religious bondage, the moderns could criticize the obscurantism of the old powers by revealing the material causality that those powers dissimulated – even as they invented those very phenomena in the artificial enclosure of the laboratory. The Laws of Nature allowed the first Enlightenment thinkers to demolish the ill-founded pretensions of human prejudice. Applying this new critical tool, they no longer saw anything in the hybrids of old but illegitimate mixtures that they had to purify by separating natural mechanisms from human passions, interests or ignorance. All the ideas of yesteryear, one after the other, became inept or approximate. Or rather, simply applying the modern Constitution was enough to create, by contrast, a 'yesteryear' absolutely different from today. The obscurity of the olden days, which illegitimately blended together social needs and natural reality, meanings and mechanisms, signs and things, gave way to a luminous dawn that cleanly separated material causality from human fantasy. The natural sciences at last defined what Nature was, and each new emerging scientific discipline was experienced as a total revolution by means of which it was finally liberated from its prescientific past, from its Old Regime. No one who has not felt the beauty of this dawn and thrilled to its promises is modern.

But the modern critique did not simply turn to Nature in order to destroy human prejudices. It soon began to move in the other direction, turning to the newly founded social sciences in order to destroy the excesses of naturalization. This was the second Enlightenment, that of the nineteenth century. This time, precise knowledge of society and its laws made it possible to criticize not only the biases of ordinary obscurantism but also the new biases created by the natural sciences. With solid support from the social sciences, it became possible to distinguish the truly scientific component of the other sciences from the component attributable to ideology. Sorting out the kernels of science from the chaff of ideology became the task for generations of well-meaning modernizers. In the hybrids of the first Enlightenment thinkers, the second group too often saw an unacceptable blend that needed to be purified by carefully separating the part that belonged to things themselves and the part that could be attributed to the functioning of the economy, the unconscious, language, or symbols. All the ideas of yesteryear – including those of certain pseudo-sciences – became inept or approximate. Or rather, by contrast, a succession of radical revolutions created an obscure 'yesteryear' that was soon to be dissipated by the luminous dawn of the social sciences. The traps of naturalization and scientific ideology were finally dispelled. No one who has not waited for that dawn and thrilled to its promises is modern.

The invincible moderns even found themselves able to combine the two critical moves by using the natural sciences to debunk the false pretensions of power and using the certainties of the human sciences to uncover the false pretensions of the natural sciences, and of scientism. Total knowledge was finally within reach. If it seemed impossible, for so long, to get past Marxism, this was because Marxism interwove the two most powerful resources ever developed for the modern critique, and bound them together for all time (Althusser, 1992). Marxism made it possible to retain the portion of truth belonging to the natural and social sciences even while it carefully eliminated their condemned portion, their ideology. Marxism realized – and finished off, as was soon to become clear – all the hopes of the first Enlightenment, along with all those of the second. The first distinction between material causality and the illusions of obscurantism, like the second distinction between science and ideology, still remain the two principal sources of modern indignation today, even though our contemporaries can no longer close off discussion in Marxist fashion, and even though their critical capital has now been disseminated into the hands of millions of small shareholders. Anyone who has never felt this dual power vibrate within, anyone who has never been obsessed by the distinction between rationality and obscurantism, between false ideology and true science, has never been modern.

Anchor point	Critical possibility
Transcendence of nature	We can do nothing against Nature's laws
Immanence of Nature	We have unlimited possibilities
Immanence of Society	We are totally free
Transcendence of Society	We can do nothing against Society's laws

Figure 2.2 Anchor points and critical possibilities

Solidly grounded in the transcendental certainty of nature's laws, the modern man or woman can criticize and unveil, denounce and express indignation at irrational beliefs and unjustified dominations. Solidly grounded in the certainty that humans make their own destiny, the modern man or woman can criticize and unveil, express indignation at and denounce irrational beliefs, the biases of ideologies, and the unjustified domination of the experts who claim to have staked out the limits of action and freedom. The exclusive transcendence of a Nature that is not our doing, and the exclusive immanence of a Society that we create through and through, would nevertheless paralyze the moderns, who would appear too impotent in the face of things and too powerful within society. What an enormous advantage to be able to reverse the principles without even the appearance of contradiction! In spite of its transcendence, Nature remains mobilizable, humanizable, socializable. Every day, laboratories, collections, centres of calculation and of profit, research bureaus and scientific institutions blend it with the multiple destinies of social groups. Conversely, even though we construct Society through and through, it lasts, it surpasses us, it dominates us, it has its own laws, it is as transcendent as Nature. For every day, laboratories, collections, centres of calculation and of profit, research bureaus and scientific institutions stake out the limits to the freedom of social groups, and transform human relations into durable objects that no one has made. The critical power of the moderns lies in this double language: they can mobilize Nature at the heart of social relationships, even as they leave Nature infinitely remote from human beings; they are free to make and unmake their society, even as they render its laws ineluctable, necessary and absolute.

2.11 The Invincibility of the Moderns

Because it believes in the total separation of humans and nonhumans, and because it simultaneously cancels out this separation, the Constitution has made the moderns invincible. If you criticize them by saying that Nature is a world constructed by human hands, they will show you that it is transcendent, that science is a mere intermediary allowing access to Nature, and that they keep their hands off. If you tell them that we are free and that our destiny is in our own hands, they will tell you that Society is transcendent and its laws infinitely surpass us. If you object that they are being duplicitous, they will show you that they never confuse the Laws of Nature with imprescriptible human freedom. If you believe them and direct your attention elsewhere, they will take advantage of this to transfer thousands of objects from Nature into the social body while procuring for this body the solidity of natural things. If you turn round suddenly, as in the children's game 'Mother, may I?', they will freeze, looking innocent, as if they hadn't budged: here, on the left, are things themselves; there, on the right, is the free society of speaking, thinking subjects, values and of signs. Everything happens in the middle, everything passes between the two, everything happens by way of mediation, translation and networks, but this space does not exist, it has no place. It is the unthinkable, the unconscious of the moderns. What

better way to extend collectives than by bringing them into alliance both
with Nature's transcendence and with all of human freedom, while at the
same time incorporating Nature and imposing absolute limits on the
boundaries of freedom? This makes it possible to do anything – and its
opposite.

Native Americans were not mistaken when they accused the Whites of
having forked tongues. By separating the relations of political power
from the relations of scientific reasoning while continuing to shore up
power with reason and reason with power, the moderns have always had
two irons in the fire. They have become invincible.

You think that thunder is a divinity? The modern critique will show
that it is generated by mere physical mechanisms that have no influence
over the progress of human affairs. You are stuck in a traditional
economy? The modern critique will show you that physical mechanisms
can upset the progress of human affairs by mobilizing huge productive
forces. You think that the spirits of the ancestors hold you forever
hostage to their laws? The modern critique will show you that you are
hostage to yourselves and that the spiritual world is your own human –
too human – construction. You then think that you can do everything
and develop your societies as you see fit? The modern critique will show
you that the iron laws of society and economics are much more inflexible
than those of your ancestors. You are indignant that the world is being
mechanized? The modern critique will tell you about the creator God to
whom everything belongs and who gave man everything. You are
indignant that society is secular? The modern critique will show you that
spirituality is thereby liberated, and that a wholly spiritual religion is far
superior. You call yourself religious? The modern critique will have a
hearty laugh at your expense!

How could the other cultures-natures have resisted? They became
premodern by contrast. They could have stood up against transcendent
Nature, or immanent Nature, or society made by human hands, or
transcendent Society, or a remote God, or an intimate God, but how
could they resist the combination of all six? Or rather, they might have
resisted, if the six resources of the modern critique had been visible
together in a single operation such as I am retracing today. But they
seemed to be separate, in conflict with one another, blending incompatible branches of government, each one appealing to different
foundations. What is more, all these critical resources of purification
were contradicted at once by the practice of mediation, yet that
contradiction had no influence whatsoever either on the diversity of the
sources of power or on their hidden unity.

Such a superiority, such an originality, made the moderns think they
were free from the ultimate restrictions that might limit their expansion.

Century after century, colonial empire after colonial empire, the poor premodern collectives were accused of making a horrible mishmash of things and humans, of objects and signs, while their accusers finally separated them totally – to remix them at once on a scale unknown until now. . . . As the moderns also extended this Great Divide in time after extending it in space, they felt themselves absolutely free to give up following the ridiculous constraints of their past which required them to take into account the delicate web of relations between things and people. But at the same time they were taking into account many more things and many more people. . .

You cannot even accuse them of being nonbelievers. If you tell them they are atheists, they will speak to you of an all-powerful God who is infinitely remote in the great beyond. If you say that this crossed-out God is something of a foreigner, they will tell you that He speaks in the privacy of the heart, and that despite their sciences and their politics they have never stopped being moral and devout. If you express astonishment at a religion that has no influence either on the way the world goes or on the direction of society, they will tell you that it sits in judgement on both. If you ask to read those judgements, they will object that religion infinitely surpasses science and politics and it does not influence them, or that religion is a social construct, or the effect of neurons!

What will you tell them, then? They hold all the sources of power, all the critical possibilities, but they displace them from case to case with such rapidity that they can never be caught redhanded. Yes, unquestionably, they are, they have been, they have almost been, they have believed they were, invincible.

2.12 What the Constitution Clarifies and What It Obscures

Yet the modern world has never happened, in the sense that it has never functioned according to the rules of its official Constitution alone: it has never separated the three regions of Being I have mentioned and appealed individually to the six resources of the modern critique. The practice of translation has always been different from the practices of purification. Or rather, this difference itself is inscribed in the Constitution, since the double play of each of the three agencies between immanence and transcendence makes it possible to do anything – and its opposite. Never has a Constitution allowed such a margin for manœuvre in practice. But the price the moderns paid for this freedom was that they remained unable to conceptualize themselves in continuity with the premoderns. They had to think of themselves as absolutely different, they had to invent the Great Divide because the entire work of mediation escapes the constitutional framework that simultaneously outlines it and denies its existence.

Expressed in this way, the modern predicament looks like a plot that I am about to unveil. False consciousness would force the moderns to imagine a Constitution that they can never apply. They would practise the very things that they are not allowed to say. The modern world would thus be populated by liars and cheaters. Worse still, by proposing to debunk their illusions, to uncover their real practice, to probe their unconscious belief, to reveal their double talk, I would play a very modern role indeed, taking my turn in a long queue of debunkers and critics. But the relation between the work of purification and that of mediation is not that of conscious and unconscious, formal and informal, language and practice, illusion and reality. I am not claiming that the moderns are unaware of what they do, I am simply saying that what they do – innovate on a large scale in the production of hybrids – is possible only because they steadfastly hold to the absolute dichotomy between the order of Nature and that of Society, a dichotomy which is itself possible only because they never consider the work of purification and that of mediation together. There is no false consciousness involved, since the moderns are explicit about the two tasks. They have to practise the top and the bottom halves of the modern Constitution. The only thing I add is the relation between those two different sets of practices.

So is modernity an illusion? No, it is much more than an illusion and much less than an essence. It is a force added to others that for a long time it had the power to represent, to accelerate, or to summarize – a power that it no longer entirely holds. The revision I am proposing is similar to the revision of the French Revolution that has been undertaken during the last twenty years or so in France – and the two revisions amount to one and the same, as we shall see further on. Since the 1970s, French historians have finally understood that the revolutionary reading of the French Revolution had been added to the events of that time, that it had organized historiography since 1789, but that it no longer defines the events themselves (Furet, [1978] 1981). As François Furet proposes, the Revolution as 'modality of historical action' is to be distinguished from the Revolution as 'process'. The events of 1789 were no more revolutionary than the modern world has been modern. The actors and chroniclers of 1789 used the notion of revolution to understand what was happening to them, and to influence their own fate. Similarly, the modern Constitution exists and indeed acts in history, but it no longer defines what has happened to us. Modernity still awaits its Tocqueville, and the scientific revolutions still await their François Furet.

So, modernity is not the false consciousness of moderns, and we have to be very careful to grant the Constitution, like the idea of Revolution, its own effectiveness. Far from eliminating the work of mediation, it has allowed this work to expand. Just as the idea of Revolution led the revolutionaries to take irreversible decisions that they would not have dared take without it, the Constitution provided the moderns with the daring to mobilize things and people on a scale that they would otherwise have disallowed. This modification of scale was achieved not – as they thought – by the separation of humans and nonhumans but, on the contrary, by the amplification of their contacts. This growth is in turn facilitated by the idea of transcendent Nature (provided that it remains mobilizable), by the idea of free Society (provided that it remains transcendent), and by the absence of all divinity (provided that God speaks to the heart). So long as their contraries remain simultaneously present and unthinkable, and so long as the work of mediation multiplies hybrids, these three ideas make it possible to capitalize on a large scale. The moderns think they have succeeded in such an expansion only because they have carefully separated Nature and Society (and bracketed God), whereas they have succeeded only because they have mixed together much greater masses of humans and nonhumans, without bracketing anything and without ruling out any combination! The link between the work of purification and the work of mediation has given birth to the moderns, but they credit only the former with their success. In saying this I am not unveiling a practice hidden beneath an official reading, I am simply adding the bottom half to the upper half. They are both necessary together, but as long as we were modern, they simply could not appear as one single and coherent configuration.

So are the moderns aware of what they are doing or not? The solution to the paradox may not be too hard to find if we look at what anthropologists tell us of the premoderns. To undertake hybridization, it is always necessary to believe that it has no serious consequences for the constitutional order. There are two ways of taking this precaution. The first consists in thoroughly thinking through the close connections between the social and the natural order so that no dangerous hybrid will be introduced carelessly. The second one consists in bracketing off entirely the work of hybridization on the one hand and the dual social and natural order on the other. While the moderns insure themselves by not thinking at all about the consequences of their innovations for the social order, the premoderns – if we are to believe the anthropologists – dwell endlessly and obsessively on those connections between nature and culture. To put it crudely: those who think the most about hybrids circumscribe them as much as possible, whereas those who choose to ignore them by insulating them from any dangerous consequences develop them to the utmost. The premoderns are all monists in the constitution of their nature-cultures. 'The native is a logical hoarder', writes Claude Lévi-Strauss; 'he is forever tying the threads, unceasingly turning over all the aspects of reality, whether physical, social or mental' (Lévi-Strauss, [1962] 1966, p. 267). By saturating the mixes of divine, human and natural elements with concepts, the premoderns limit the practical expansion of these mixes. It is the impossibility of changing the social order without modifying the natural order – and vice versa – that has obliged the premoderns to exercise the greatest prudence. Every monster becomes visible and thinkable and explicitly poses serious problems for the social order, the cosmos, or divine laws (Horton, 1967, 1982). Descola writes about the Achuar:

> The homeostasis of the 'cold societies' of Amazonia would be less the result of the implicit rejection of political alienation, with which Clastres credited 'savages' (Clastres, 1974) . . . than the effect of the inertia effect of a thought system *unable to represent the process of socializing nature in any way other than through the categories that dictate the way real society should function.* Running counter to the overhasty technical determinism with which evolutionist theories are often imbued, one might postulate that when a society transforms its material base, this is conditioned by a prior mutation of the forms of social organization that comprise the conceptual framework of the material mode of producing. (Descola, [1986] 1993; p. 405; emphasis added)

If, on the contrary, our Constitution authorizes anything, it is surely the accelerated socialization of nonhumans, because it never allows them to appear as elements of 'real society'. By rendering mixtures unthinkable, by emptying, sweeping, cleaning and purifying the arena that is opened in the central space defined by their three sources of power, the moderns allowed the practice of mediation to recombine all possible monsters without letting them have any effect on the social fabric, or even any contact with it. Bizarre as these monsters may be, they posed no problem because they did not exist publicly and because their monstrous consequences remained untraceable. What the premoderns have always ruled out the moderns can allow, since the social order never turns out to correspond, point for point, with the natural order.

Boyle's air pump, for example, might seem to be a rather frightening chimera, since it produces a laboratory vacuum artificially, a vacuum that simultaneously permits the definition of the Laws of Nature, the action of God, and the settlement of disputes in England at the time of the Glorious Revolution. According to Robin Horton, savage thought would have conjured away its dangers at once. From now on the English seventeenth century will go on to construct Royalty, Nature and theology with the scientific community and the laboratory. The air's spring will join the actors that inhabit England. Yet this recruitment of a new ally poses no problem, since there is no chimera, since nothing monstrous has been produced, since nothing more has been done than to discover the Laws of Nature. *The scope of the mobilization is directly proportional to the impossibility of directly conceptualizing its relations with the social order.* The less the moderns think they are blended, the more they blend. The more science is absolutely pure, the more it is intimately bound up with the fabric of society. The modern Constitution accelerates or facilitates the deployment of collectives – which differ, as I indicated earlier, from societies made up only of social relations – but does not allow their conceptualization.

2.13 The End of Denunciation

To be sure, by affirming that the Constitution, if it is to be effective, has to be unaware of what it allows, I am practising an unveiling, but one that no longer bears upon the same objects as the modern critique and is no longer triggered by the same mainsprings. So long as we adhered willingly to the Constitution, it allowed us to settle all disputes and served as a basis for the critical spirit, providing individuals with justification for their attacks and their operations of unveiling. But if the Constitution as a whole now appears as only one half that no longer allows us to understand its own other half, then it is the very foundation of the modern critique that turns out to be ill-assured. I am thus trying the tricky move to unveil the modern Constitution without resorting to the modern type of debunking. To do so I am accounting for this vague and uneasy feeling that we have recently become as unable to denounce as to modernize. The upper ground for taking a critical stance seems to have escaped us.

Yet by appealing sometimes to Nature, sometimes to Society, sometimes to God, and by constantly opposing the transcendence of each one of these three terms to its immanence, the moderns had found the mainspring of their indignations well wound up. What kind of a modern could no longer fall back on the transcendence of nature to criticize the obscurantism of power? On the immanence of Nature to criticize human inertia? On the immanence of Society to criticize the submission of humans and the dangers of naturalism? On the transcendence of society to criticize the human illusion of individual liberty? On the transcendence of God to appeal to the judgement of humans and the obstinacy of things? On the immanence of God to criticize established Churches, naturalist beliefs and socialist dreams? It would be a pretty pathetic kind of modern, or else a postmodern: still inhabited by the violent desire to denounce, they would no longer have the strength to believe in the legitimacy of any of these six courts of appeal. To strip moderns of their indignation is to deprive them, it seems, of all self-respect. To strip critical intellectuals of the six bases for their denunciations is apparently to rob them of all reason to live. In losing our wholehearted adherence to the Constitution, do we not have the impression that we are losing the best of ourselves? Was it not the origin of our energy, our moral strength, our ethics?

However, Luc Boltanski and Laurent Thévenot have done away with modern denunciation, in a book as important for my own essay as Shapin and Schaffer's. They have done for the work of critical indignation what François Furet did earlier for the French Revolution. 'The French Revolution is over,' he wrote; in the same vein the subtitle of *Économies de la grandeur* could have been 'The modern denunciation is over' (Boltanski and Thévenot, 1991). Up to that point, critical unmasking appeared to be self-evident. It was only a matter of choosing a cause for indignation and opposing false denunciations with as much passion as possible. To unmask: that was our sacred task, the task of us moderns. To reveal the true calculations underlying the false conscious-nesses, or the true interests underlying the false calculations. Who is not still foaming slightly at the mouth with that particular rabies? Now Boltanski and Thévenot have invented the equivalent of an anti-rabies vaccine by calmly comparing all sources of denunciation – the Cities that supply the various principles of justice – and by interweaving the thousand and one ways we have, in France today, of bringing an affair to justice. They do not denounce others. They do not unmask anyone. They show how we all go about accusing one another. Instead of a resource, the critical spirit becomes a topic, one competence among others, the grammar of our indignations. Instead of practising a critical sociology the authors quietly begin a sociology of criticism.

Suddenly, thanks to this little gap opened up by systematic study, we can no longer fully adhere to the spirit of the modern critique. How can we still make wholehearted accusations when the scapegoating mechan-ism has become obvious? Even the human sciences are no longer the ultimate reservoir that would make it possible at last to discern the real motives beneath appearances. They too are made part of the analysis (Chateauraynaud, 1990); they too bring issues to justice, and become indignant and criticize. The tradition of the human sciences no longer has the privilege of rising above the actor by discerning, beneath his unconscious actions, the reality that is to be brought to light (Boltanski, 1990). It is impossible for the human sciences to be scandalized, without henceforth occupying one of the boxes in our colleagues' grid. The denouncer is the brother of the ordinary people that he claimed to be denouncing. Instead of really believing in it, we now experience the work of denunciation as a 'historical modality' which certainly influences our affairs but does not explain them any more than the revolutionary modality explained the process of the events of 1789. Today, denunciation and revolution have both gone stale.

Boltanski and Thévenot's work completes the movement predicted and described by René Girard according to which moderns can no longer make sincere accusations; but Boltanski and Thévenot, unlike Girard, do not scorn objects. In order for the mechanism of victim-formation to function, the accused person who was sacrificed in public by the crowd had to be actually guilty (Girard, [1978] 1987). If the victim became a scapegoat, the mechanism of accusation became visible: some fall guy innocent of any crime was wrongly accused, with no reason except to reconcile the community at his expense. The shift from sacrifice to scapegoat thus voids accusation. This evacuation does not soften the moderns, however, since the reason for their series of crimes is precisely that they are never able to make a genuine accusation of a truly guilty party (Girard, 1983). But Girard does not see that he himself is thus making a more serious allegation, since he accuses objects of not really counting. So long as we imagine objective stakes for our disputes, he claims, we are caught up in the illusion of mimetic desire. It is this desire, and this desire alone, that adorns objects with a value that is not their own. In themselves, they do not count; they are nothing. By revealing the process of accusation, Girard, like Boltanski and Thévenot, forever exhausts our aptitude to accuse. But he prolongs the tendency of moderns to scorn objects even further – and Girard tenders that accusation wholeheartedly; he really believes it, and he sees in this hard-won scorn the highest proof of morality (Girard, 1989). Here is a denouncer and a half. The greatness of Boltanski and Thévenot's book comes from the fact that they exhaust denunciation even as they put the object engaged in tests of judgement at the heart of their analyses.

Are we devoid of any moral foundation once denunciation has been exhausted? But underneath moral judgement by denunciation, another moral judgement has always functioned by triage and selection. It is called arrangement, combination, *combinazione*, combine, but also negotiation or compromise. Charles Péguy used to say that a supple morality is infinitely more exigent than a rigid morality (Péguy, 1961b). The same holds true for the unofficial morality that constantly selects and distributes the practical solutions of the moderns. It is scorned because it does not allow indignation, but it is active and generous because it follows the countless meanderings of situations and networks. It is scorned because it takes into account the objects that are no more the arbitrary stakes of our desire alone than they are the simple receptacle for our mental categories. Just as the modern Constitution scorns the hybrids that it shelters, official morality scorns practical arrangements and the objects that uphold it. Underneath the opposition between objects and subjects, there is the whirlwind of the mediators. Underneath moral grandeur there is the meticulous triage of circumstances and cases (Jonsen and Toulmin, 1988).

2.14 We Have Never Been Modern

I now have a choice: either I believe in the complete separation between the two halves of the modern Constitution, or I study both what this Constitution allows and what it forbids, what it clarifies and what it obfuscates. Either I defend the work of purification – and I myself serve as a purifier and a vigilant guardian of the Constitution – or else I study both the work of mediation and that of purification – but I then cease to be wholly modern.

By claiming that the modern Constitution does not permit itself to be understood, by proposing to reveal the practices that allow it to exist, by asserting that the critical mechanism has outlived its usefulness, am I behaving as though we were entering a new era that would follow the era of the moderns? Would I then be, literally, postmodern? Postmodernism is a symptom, not a fresh solution. It lives under the modern Constitution, but it no longer believes in the guarantees the Constitution offers. It senses that something has gone awry in the modern critique, but it is not able to do anything but prolong that critique, though without believing in its foundations (Lyotard, 1979). Instead of moving on to empirical studies of the networks that give meaning to the work of purification it denounces, postmodernism rejects all empirical work as illusory and deceptively scientistic (Baudrillard, 1992). Disappointed rationalists, its adepts indeed sense that modernism is done for, but they continue to accept its way of dividing up time; thus they can divide up eras only in terms of successive revolutions. They feel that they come 'after' the moderns, but with the disagreeable sentiment that there is no more 'after'. 'No future': this is the slogan added to the moderns' motto 'No past'. What remains? Disconnected instants and groundless denunciations, since the postmoderns no longer believe in the reasons that would allow them to denounce and to become indignant.

A different solution appears as soon as we follow both the official Constitution and what it forbids or allows, as soon as we study in detail the work of production of hybrids and the work of elimination of these same hybrids. We then discover that we have never been modern in the sense of the Constitution, and this is why I am not debunking the false consciousness of people who would practise the contrary of what they claim. No one has ever been modern. Modernity has never begun. There has never been a modern world. The use of the past perfect tense is important here, for it is a matter of a retrospective sentiment, of a rereading of our history. I am not saying that we are entering a new era; on the contrary we no longer have to continue the headlong flight of the post-post-postmodernists; we are no longer obliged to cling to the avant-garde of the avant-garde; we no longer seek to be even cleverer, even more critical, even deeper into the 'era of suspicion'. No, instead we discover that we have never begun to enter the modern era. Hence the hint of the ludicrous that always accompanies postmodern thinkers; they claim to come after a time that has not even started!

This retrospective attitude, which deploys instead of unveiling, adds instead of subtracting, fraternizes instead of denouncing, sorts out instead of debunking, I characterize as nonmodern (or amodern). A nonmodern is anyone who takes simultaneously into account the moderns' Constitution and the populations of hybrids that that Constitution rejects and allows to proliferate.

The Constitution explained everything, but only by leaving out what was in the middle. 'It's nothing, nothing at all,' it said of the networks, 'merely residue.' Now hybrids, monsters – what Donna Haraway calls 'cyborgs' and 'tricksters' (Haraway, 1991) whose explanation it abandons – are just about everything; they compose not only our own collectives but also the others, illegitimately called premodern. At the very moment when the twin Enlightenments of Marxism seemed to have explained everything, at the very moment when the failure of their total explanation leads the postmoderns to founder in the despair of self-criticism, we discover that the explanations had not yet begun, and that this has always been the case; that we have never been modern, or critical; that there has never been a yesteryear or an Old Regime (Mayer, 1982); that we have never really left the old anthropological matrix behind, and that it could not have been otherwise.

To notice that we have never been modern and that only minor divisions separate us from other collectives does not mean that I am a reactionary. The antimodern reaction struggles fiercely against the effects of the Constitution, but accepts it fully. Antimoderns want to defend localities, or spirit, or rationality, or the past, or universality, or liberty, or society, or God, as if these entities really existed and actually had the form that the official part of the modern Constitution granted them. Only the sign and the direction of their indignation vary. The antimoderns even accept the chief oddity of the moderns, the idea of a time that passes irreversibly and annuls the entire past in its wake. Whether one wishes to conserve such a past or abolish it, in either case the revolutionary idea *par excellence*, the idea that revolution is possible, is maintained. Today, that very idea strikes us as exaggerated, since revolution is only one resource among many others in histories that have nothing revolutionary, nothing irreversible, about them. '*In potentia*' the modern world is a total and irreversible invention that breaks with the past, just as '*in potentia*' the French or Bolshevik Revolutions were midwives at the birth of a new world. Seen as networks, however, the modern world, like revolutions, permits scarcely anything more than small extensions of practices, slight accelerations in the circulation of knowledge, a tiny extension of societies, minuscule increases in the number of actors, small modifications of old beliefs. When we see them as networks, Western innovations remain recognizable and important, but they no longer suffice as the stuff of saga, a vast saga of radical rupture, fatal destiny, irreversible good or bad fortune.

The antimoderns, like the postmoderns, have accepted their adversaries' playing field. Another field – much broader, much less polemical – has opened up before us: the field of nonmodern worlds. It is the Middle Kingdom, as vast as China and as little known.

3

□

REVOLUTION

3.1 The Moderns, Victims of Their Own Success

If the critical apparatus of the moderns has made them invincible, why are they hesitating over their own destiny today? If the effectiveness of the Constitution depended precisely upon its obscure half, why can I now relate it to its luminous half? The bond between the two sets of practices must indeed have changed for me to be able to follow both the practices of purification and those of translation. If we can no longer adhere wholeheartedly to the tasks of modernization, unforeseen obstacles must have interfered with the mechanism. What has happened that makes the work of purification unthinkable, when a few years ago it was the deployment of networks that appeared absurd and scandalous?

Let us say that the moderns have been victims of their own success. It is a crude explanation, I admit, yet it would appear that the scope of the mobilization of collectives had ended up multiplying hybrids to such an extent that the constitutional framework which both denies and permits their existence could no longer keep them in place. The modern Constitution has collapsed under its own weight, submerged by the mixtures that it tolerated as material for experimentation because it simultaneously dissimulated their impact upon the fabric of society. The third estate ends up being too numerous to feel that it is faithfully represented either by the order of objects or by the order of subjects.

When the only thing at stake was the emergence of a few vacuum pumps, they could still be subsumed under two classes, that of natural laws and that of political representations; but when we find ourselves invaded by frozen embryos, expert systems, digital machines, sensor-equipped robots, hybrid corn, data banks, psychotropic drugs, whales outfitted with radar sounding devices, gene synthesizers, audience analyzers, and so on, when our daily newspapers display all these monsters on page after page, and when none of these chimera can be properly on the object side or on the subject side, or even in between, something has to be done. It is as if the two poles of the Constitution had been conflated in the end precisely because of the practice of mediation that this Constitution at once liberates and disavows. It is as if there were no longer enough judges and critics to partition the hybrids. The purification system has become as clogged as our judicial system.

Perhaps the modern framework could have held up a little while longer if its very development had not established a short circuit between Nature on the one hand and human masses on the other. So long as Nature was remote and under control, it still vaguely resembled the constitutional pole of tradition, and science could still be seen as a mere intermediary to uncover it. Nature seemed to be held in reserve, transcendent, inexhaustible, distant enough. But where are we to classify the ozone hole story, or global warming or deforestation? Where are we to put these hybrids? Are they human? Human because they are our work. Are they natural? Natural because they are not our doing. Are they local or global? Both. As for the human masses that have been made to multiply as a result of the virtues and vices of medicine and economics, they are no easier to situate. In what world are these multitudes to be housed? Are we in the realm of biology, sociology, natural history, ethics, sociobiology? This is our own doing, yet the laws of demography and economics are infinitely beyond us. Is the demographic time bomb local or global? Both. Thus, the two constitutional guarantees of the moderns – the universal laws of things, and the inalienable rights of subjects – can no longer be recognized either on the side of Nature or on the side of the Social. The destiny of the starving multitudes and the fate of our poor planet are connected by the same Gordian knot that no Alexander will ever again manage to sever.

Let us say, then, that the moderns have caved in. Their Constitution could absorb a few counter-examples, a few exceptions – indeed, it thrived on them. But it is helpless when the exceptions proliferate, when the third estate of things and the Third World join together to invade all its assemblies *en masse*. In order to accommodate those exceptions, which are hardly any different from those of savage thought (see below), we need to outline a space that is no longer the space of the modern Constitution, because it fills the median zone that the Constitution claimed to empty. To the practice of purification – the horizontal line – we need to add the practices of mediation – the vertical line.

Instead of following the multiplication of hybrids by projecting them on to their longitude alone, we also need to identify them by means of a latitude. The diagnosis of the crisis with which I began this essay is now

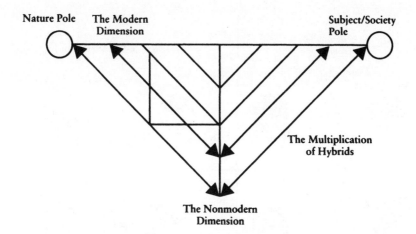

Figure 3.1 Purification and mediation

quite clear: *the proliferation of hybrids has saturated the constitutional framework of the moderns.* The moderns have always been using both dimensions in practice, they have always been explicit about each of them, but they have never been explicit about the relation between the two sets of practices. Nonmoderns have to stress the relations beteen them if they are to understand both the moderns' successes and their recent failures, and still not lapse into postmodernism. By deploying both dimensions at once, we may be able to accommodate the hybrids and give them a place, a name, a home, a philosophy, an ontology and, I hope, a new constitution.

3.2 What Is a Quasi-Object?

Using the two dimensions at once, the longitude and the latitude, we may now be able to locate the position of these strange new hybrids and to understand how come that we had to wait for science studies in order to define what, following Michel Serres (1987), I shall call quasi-objects, quasi-subjects. To do so, we simply have to follow the little comic strip in Figure 3.2.

Social scientists have for long allowed themselves to denounce the belief system of ordinary people. They call this belief system 'naturalization' (Bourdieu and Wacquant, 1992). Ordinary people imagine that the power of gods, the objectivity of money, the attraction of fashion, the beauty of art, come from some objective properties intrinsic to the nature of things. Fortunately, social scientists know better and they show that

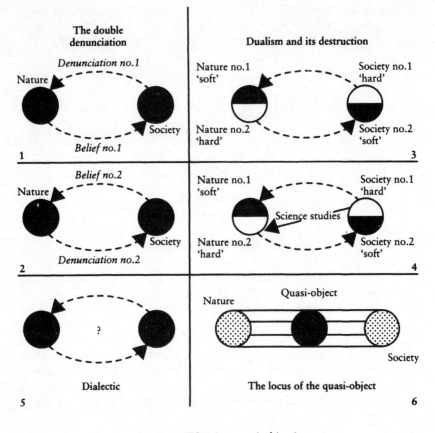

Figure 3.2 What is a quasi-object?

the arrow goes in fact in the other direction, from society to the objects. Gods, money, fashion and art offer only a surface for the projection of our social needs and interests. At least since Emile Durkheim, such has been the price of entry into the sociology profession (Durkheim, [1915] 1965). To become a social scientist is to realize that the inner properties of objects do not count, that they are mere receptacles for human categories.

The difficulty, however, is to reconcile this form of denunciation with another one in which the directions of the arrows are exactly reversed. Ordinary people, mere social actors, average citizens, believe that they are free and that they can modify their desires, their motives and their rational strategies at will. The arrow of their beliefs now goes from the Subject/Society pole to the Nature pole. But fortunately, social scientists are standing guard, and they denounce, and debunk and ridicule this naive belief in the freedom of the human subject and society. This time they use the nature of things – that is the indisputable results of the sciences – to show how it determines, informs and moulds the soft and pliable wills of the poor humans. 'Naturalization' is no longer a bad word but the shibboleth that allows the social scientists to ally themselves with the natural sciences. All the sciences (natural and social) are now mobilized to turn the humans into so many puppets manipulated by objective forces – which only the natural or social scientists happen to know.

When the two critical resources are put together we now understand why it is so difficult for social scientists to reach agreement on objects. They too 'see double'. In the first denunciation, objects count for nothing; they are just there to be used as the white screen on to which society projects its cinema. But in the second, they are so powerful that they shape the human society, while the social construction of the sciences that have produced them remains invisible. Objects, things, consumer goods, works of art are either too weak or too strong. But still stranger are the successive roles given to society. In the first denunciation, society is so powerful that it is *sui generis*, it has no more cause than the transcendental ego it replaces. It is so originary that it is able to mould and shape what is nothing more than an arbitrary and shapeless matter. In the second form of denunciation, however, it has become powerless, shaped in turn by the powerful objective forces that completely determine its action. Society is either too powerful or too weak *vis-à-vis* objects which are alternatively too powerful or too arbitrary.

The solution to this double contradictory denunciation is so pervasive that it has been providing social scientists with most of their common sense; it is called dualism. The Nature pole will be partitioned into two sets: the first list will incude its 'softer' parts – screens for projecting social categories – while the second list will include all its 'harder' parts – causes for determining the fate of human categories: that is, the sciences and the technologies. The same partition will be made on the Subject/ Society pole: there will be its 'harder' components – the *sui generis* social factors – and its 'softer' components – determined by the forces discovered by sciences and technologies. Social scientists will happily alternate from one to the other showing without any trouble that for instance gods are mere idols shaped by the requirements of social order, while the rules of society are determined by biology.

To be sure, this alternation is not very convincing. First, the lists are made haphazardly, the 'soft' list of the nature pole gathering all the things social scientists happen to despise – religion, consumption, popular culture and politics – while the 'hard' list is made of all the sciences they naively believe in at the time – economics, genetics, biology, linguistics, or brain sciences. Second, it is not clear why society needs to be projected on to arbitrary objects if those objects count for nothing. Is society so weak that it needs continuous resuscitation? So terrible that, like Medusa's face, it should be seen only in a mirror? And if religion, arts or styles are necessary to 'reflect', 'reify', 'materialize', 'embody' society – to use some of the social theorists' favourite verbs – then are objects not, in the end, its co-producers? Is not society built literally – not metaphorically – of gods, machines, sciences, arts and styles? But then where is the illusion of the 'common' actor in the bottom arrow of Figure 3.2.1? Maybe social scientists have simply forgotten that before projecting itself on to things society has to be made, built, constructed? And out of what material could it be built if not out of nonsocial, non-human resources? But social theory is forbidden to draw this conclusion because it has no conception of objects except the one handed down to it by the alternative 'hard' sciences which are so strong that they simply determine social order which in turn becomes flimsy and immaterial.

Dualism may be a poor solution, but it provided 99 per cent of the social sciences' critical repertoire, and nothing would have disturbed its blissful asymmetry if science studies had not upset the applecart. Up to that point, dualism had seemed to work, since the 'hard' part of society was used on the 'soft' objects, while the 'hard' objects were used only on the 'soft' part of society (Bourdieu and Wacquant, 1992). Social scientists could denounce the practices they did not believe in by using the solid science of society they had concocted and embracing the sciences they had complete confidence in so as to establish the social order. It is the glory of the Edinburgh school of social studies of science to have attempted a forbidden crossover (Barnes, 1974; Barnes and Shapin, 1979; Bloor, [1976] 1991; MacKenzie, 1981; Shapin, 1992). They used the critical repertoire that was reserved for the 'soft' parts of nature to debunk the 'harder' parts, the sciences themselves! In short, they wanted to do for science what Durkheim had done for religion, or Bourdieu for fashion and taste; and they innocently thought that the social sciences would remain unchanged, swallowing science as easily as religion or the arts. But there was a big difference, invisible until then. Social scientists did not really believe in religion and popular consumption. They did believe in science, however, from the bottom of their scientistic hearts.

Thus this breach of the dualists' game immediately bankrupted the whole enterprise. What had started as a 'social' study of science could not succeed, of course, and this is why it lasted only a split second – just long enough to reveal the terrible flaws of dualism. By treating the 'harder' parts of nature in the same way as the softer ones – that is, as arbitrary constructions determined by the interests and requirements of a *sui generis* society – the Edinburgh daredevils deprived the dualists – and indeed themselves, as they were soon to realize – of half of their resources. Society had to produce everything arbitrarily including the cosmic order, biology, chemistry, and the laws of physics! The implausibility of this claim was so blatant for the 'hard' parts of nature that we suddenly realized how implausible it was for the 'soft' ones as well. Objects are not the shapeless receptacles of social categories – neither the 'hard' ones nor the 'soft' ones. By disturbing the dualist pack of cards, the social students of science, revealed the complete asymmetry of the first and second denunciations, and they also revealed – at least negatively – how badly constructed were the social theory as well as the epistemology that went with those denunciations. Society is neither that strong nor that weak; objects are neither that weak nor that strong. The double position of objects and society had to be entirely rethought.

To resort to dialectical reasoning was no way to exit out of the difficulty into which 'science studies' had put the social sciences. Linking the two poles of nature and society by as many arrows and feedback loops as one wishes does not relocate the quasi-objects or quasi-subject that I want to take into account. On the contrary, dialectics makes the ignorance of that locus still deeper than in the dualist paradigm since it feigns to overcome it by loops and spirals and other complex acrobatic figures. Dialectics literally beats around the bush. Quasi-objects are in between and below the two poles, at the very place around which dualism and dialectics had turned endlessly without being able to come to terms with them. Quasi-objects are much more social, much more fabricated, much more collective than the 'hard' parts of nature, but they are in no way the arbitrary receptacles of a full-fledged society. On the other hand they are much more real, nonhuman and objective than those shapeless screens on which society – for unknown reasons – needed to be 'projected'. By trying the impossible task of providing social explanations for hard scientific facts – after generations of social scientists had tried either to denouce 'soft' facts or to use hard sciences uncritically – science studies have forced everyone to rethink anew the role of objects in the construction of collectives, thus challenging philosophy.

3.3 Philosophies Stretched Over the Yawning Gap

How have the major philosophies attempted to absorb both the modern Constitution and the quasi-objects, that Middle Kingdom which kept on expanding? By simplifying considerably, we can identify three principal strategies. The first consists in establishing a great gap between objects and subjects and continually increasing the distance between them; the second, known as the 'semiotic turn', focuses on the middle and abandons the extremes; the third isolates the idea of Being, thus rejecting the whole divide between objects, discourse and subjects.

Let me undertake a rapid survey of the first group. The more quasi-objects multiply, the more the major philosophies treat the two constitutional poles as incommensurable, even while they assert that there is no task more urgent than their reconciliation. So these philosophies illustrate the modern paradox in their own fashion by forbidding what they allow and allowing what they forbid. Each of these philosophies is, of course, infinitely more subtle than my inadequate summary; each one is by definition nonmodern since modernism has never really begun; thus each explicitly addresses the same problem I am awkwardly attempting to address; but their official and popularized interpretations nevertheless attest, on this point, to an astonishing consistency in the way they define their task: how to multiply quasi-objects without accepting them, in order to maintain the Great Divide that separates us both from our past and from other nature-cultures.

Hobbes and Boyle, as we have seen, fought so much only because they were just barely managing to separate the pole of natural mute nonhumans from the pole of conscious speaking citizens. The two artifacts were still so similar and so close to their common origin that the two philosophers could do no more than make a small cut through the hybrids. It is with Kantianism that our Constitution receives its truly canonical formulation. What was a mere distinction is sharpened into a total separation, a Copernican Revolution. Things-in-themselves become inaccessible while, symmetrically, the transcendental subject becomes infinitely remote from the world. The two guarantees remain clearly symmetrical, however, since knowledge is possible only at the median point, that of phenomena, through an application of the two pure forms, the thing-in-itself and the subject. Hybrids are indeed accepted, but solely as mixtures of pure forms in equal proportion. To be sure, the work of mediation remains visible, since Kant multiplies the stages needed to pass from the remote world of things to the still more remote world of the Ego. These mediations, however, are accepted only as simple inter-mediaries, which merely betray or transmit pure forms – the only recognizable ones. Multiplying layers of intermediaries make it possible to accept the role of the quasi-objects, but without giving them an ontology that would call the 'Copernican Revolution' back into question. This Kantian formulation is still visible today every time the human mind is credited with the capacity to impose forms arbitrarily on amorphous but real matter. To be sure, the Sun King around which objects revolve will be overturned in favour of many other pretenders – Society, epistemes, mental structures, cultural categories, intersubjectivity, language; but these palace revolutions will not alter the focal point, which I have called, for that reason, Subject/Society.

The greatness of dialectics derives from its attempt to traverse the complete circle of the premoderns, one last time, by encompassing all divine, social and natural beings, in order to avoid the Kantianist contradiction between the role of purification and that of mediation. But dialectics picked the wrong contradiction. It did manage to identify the one between the Subject pole and the Object pole, but it did not see the one between the whole of the modern Constitution that was establishing itself and the proliferation of quasi-objects – a proliferation that marked the nineteenth century, however, as much as it has marked our own. Or rather, dialectics thought it would absorb the second by resolving the first. Yet by believing that he was abolishing Kant's separation between things-in-themselves and the subject, Hegel brought the separation even more fully to life. He raised it to the level of a contradiction, pushed it to the limit and beyond, then made it the driving force of history. The seventeenth-century distinction becomes a separation in the eighteenth century, then an even more complete contradiction in the nineteenth. It became the mainspring of the entire plot. How could the modern paradox be better illustrated? Dialectics further enlarges the abyss that separates the Object pole from the Subject pole, but since it surmounts and abolishes this abyss in the end, it imagines that it has gone beyond Kant! Dialectics speaks of nothing but mediations, yet the countless mediations with which it peoples its grandiose history are only intermediaries that transmit pure ontological qualities – either of the spirit, in its right-wing version, or of matter, in its left-wing version. In the end, if there is a pair that no one can reconcile, it is the pole of Nature and the pole of Spirit, since their very opposition is retained and abolished – that is to say, denied. One can hardly be more modern than this. The dialecticians were incontestably our greatest modernizers, all the more powerful in that they seemed in fact to have gathered up the totality of knowledge and the past and brought to bear all the resources of the modern critique.

But quasi-objects continue to proliferate: those monsters of the first, second and third industrial revolutions, those socialized facts and these humans turned into elements of the natural world. No sooner are totalities closed in on themselves than they start cracking all over. The end of history is followed by history no matter what.

Again, one last time, phenomenology was to establish the great split, but this time with less ballast: it jettisoned the two poles of pure consciousness and pure object and spread itself, literally, over the middle, in an attempt to cover the now gaping hole that it sensed it could no

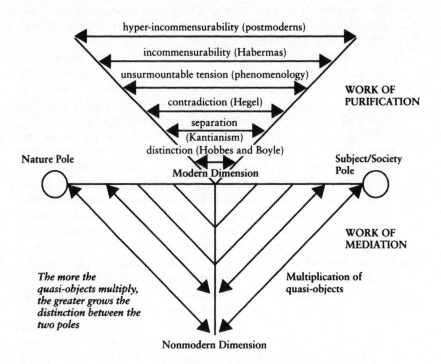

Figure 3.3 The modern paradox

longer absorb. Once again the modern paradox is taken further. The notion of intentionality transforms a distinction, a separation, a contradiction, into an insurmountable tension between object and subject. The hopes of dialectics are abandoned, since this tension offers no resolution. The phenomenologists have the impression that they have gone further than Kant and Hegel and Marx, since they no longer attribute any essence either to pure subjects or to pure objects. They really have the impression that they are speaking only of a mediation that does not require any pole to hold fast. Yet like so many anxious modernizers, they no longer trace anything but a line between poles that are thus given the greatest importance. Pure objectivity and pure consciousness are missing, but they are nevertheless – indeed, all the more – in place. The 'consciousness of something' becomes nothing more than a slender footbridge spanning a gradually widening abyss. Phenomenologists had to cave in – and they did. During the same period, Gaston Bachelard's dual enterprise – which further exaggerates the objectivity of the sciences by dint of breaking with common sense, and symmetrically exaggerates the objectless power of the imaginary by dint of epistemological breaks – offers the perfect symbol for this impossible crisis, this drawing and quartering (Bachelard, 1967; Tile, 1984).

3.4 The End of Ends

The sequel to this story takes an involuntarily comic turn. The further the great gap is stretched, the more the whole business looks like a tightrope walker doing the splits. Up to this point, all these great philosophical movements were profound and serious; they established, they explored, they accompanied the prodigious development of quasi-objects; they wanted to believe, in spite of everything, that these objects could be swallowed up and digested. By speaking only of purity, they were aiming only at grasping the work of the hybrids. All these thinkers were passionately interested in the exact sciences, in technologies and economies, because they recognized in them both the risk and the possibility of salvation. But what can be said of the philosophies that came later? And in the first place, what are we to call them? Modern? No, because they no longer attempt to hold on to both ends of the chain. Postmodern? Not yet; the worst is still to come. Let us call them pre-postmodern, to indicate that they are transitional. They raise what had been only a distinction, then a separation, then a contradiction, then an insurmountable tension, to the level of an incommensurability.

The modern Constitution as a whole had already declared that there is no common measure between the world of subjects and the world of objects, but that same Constitution at once cancelled out the distance by practising the contrary, by measuring humans and things alike with the same yardsticks, by multiplying mediators in the guise of intermediaries. The pre-postmoderns, for their part, truly believe that speaking subjects are incommensurable with natural objects and with technological efficacy, or that speaking subjects ought to become so if they are not incommensurable enough already. Thus they cancel out the modern project while claiming that they are restoring it, since they comply with the half of the Constitution that speaks of purity but neglect the other half, which practises only hybridization. They imagine that there are not – that there must not be – any mediators. On the subject side, they invent speech, hermeneutics and meaning, and they let the world of things drift slowly in its void. On the other side of the mirror, of course, scientists and technocrats take the symmetrical attitude. The more hermeneutics spins its web, the more naturalism does the same. But this repetition of the divisions of history becomes a caricature: E. O. Wilson and his genes on one side; Lacan and his analysands on the other. This pair of twins is no longer faithful to the modern intention, since they no longer make the effort to think through the paradox that consists in multiplying below the hybrids whose existence is precluded above, and in imagining impossible relations between the two.

It is worse still when the modern project is defended against the threat of disappearance. Jürgen Habermas (1987) makes one of the most desperate attempts. Is he going to show at last that nothing has ever profoundly separated things from people? Is he going to take up the modern project once again? Will he demonstrate the practical arrangements that underlie the justifications of the Constitution and finally accept the masses of hybrids as de Gaulle and Nixon finally recognized mainland China? Quite the contrary: he judges that the supreme danger arises from the confusion of speaking and thinking subjects with the pure scientific and technical rationality that is allowed by the old philosophy of consciousness! 'I have already suggested that the paradigm of the knowledge of objects has to be replaced by the paradigm of mutual understanding between subjects capable of speech and action' (p. 295–6). If anyone has ever picked the wrong enemy, it is surely this displaced twentieth-century Kantianism that attempts to widen the abyss between the objects known by the subject on the one hand, and communicational reason on the other; whereas the old consciousness had at least the merit of aiming at the object, and thus of recalling the artificial origin of the two constitutional poles. But Habermas wants to make the two poles incommensurable, at the very moment when quasi-objects are multiplying to such an extent that it appears impossible to find a single one that more or less resembles a free speaking subject or a reified natural object. Kant was already unable to bring it off in the middle of the Industrial Revolution; how could Habermas manage it after the sixth or seventh revolution? And even so, Kant multiplied the layering of intermediaries that allowed him to re-establish the transitions between things-in-themselves and the transcendental Ego. There is nothing of the sort when technological reason has to be kept as remote as possible from the free discussion of human beings.

The pre-postmoderns have something in common with the feudal reaction at the very end of the Old Regime: never was the sense of honour more prickly nor the calculation of degrees of nobility more precise; yet it was a bit late to bring off a radical separation between the third estate and the nobility! In the same way, it is a bit too late to carry off the coup of the Copernican Revolution and make things revolve around intersubjectivity. Habermas and his disciples hold on to the modern project only by abstaining from all empirical inquiry – not a single case study in the five hundred pages of his master work (Habermas, [1981] 1989); such an inquiry would bring the third estate to light too quickly, and would be too intimately mixed up with the poor speaking subjects. Let the networks perish, Habermas would say, provided that communicational reason appears to triumph.

Nevertheless, he remains honest and respectable. Even in the caricature of the modern project we can still recognize the faded splendour of the eighteenth-century Enlightenment, or the echo of the nineteenth-century Critique. Even in this obsession with separating objectivity from communication we can grasp a trace, a reminder, a scar arising from the very impossibility of bringing off such a separation. With the postmoderns, the abandonment of the modern project is consummated. I have not found words ugly enough to designate this intellectual movement – or rather, this intellectual immobility through which humans and non-humans are left to drift. I call it 'hyper-incommensurability'.

A single modern example will illustrate the abdication of thought as well as the self-inflicted defeat of the postmodern project. 'As a philosopher, I offer a balance sheet of disaster,' replies Jean-François Lyotard, who was being asked by some well-meaning scientists to conceptualize the bond that links science to the human community:

> I simply maintain that there is nothing human about scientific expansion. Perhaps our brain is only the temporary bearer of a process of complexification. It would then be a matter of detaching this process from what has supported it up to now. I am convinced that that is what you people [scientists!] are in the process of doing. Computer science, genetic engineering, physics and astrophysics, astronautics, robotics, these disciplines are already working toward preserving that complexity under conditions of life independent of life on Earth. But I do not see in what respect this is human, if by human we mean collectivities with their cultural traditions, established in a given period in precise locations on this planet. I don't doubt for a second that this 'a-human' process may have some useful fringe benefits for humanity alongside its destructive effects. But this has nothing to do with the emancipation of human beings. (Lyotard, 1988, p. xxxviii)

To the scientists who are surprised by this disastrous reckoning, and continue to believe in the usefulness of philosophers, Lyotard replies lugubriously: 'I think you have a long time to wait!' But the debacle is that of postmodernism, not that of philosophy (Hutcheon, 1989; Jameson, 1991). The postmoderns believe they are still modern because they accept the total division between the material and technological world on the one hand and the linguistic play of speaking subjects on the other – thus forgetting the bottom half of the modern Constitution; or because they relish only in the hybrid character of free floating networks and collages – thus forgetting the upper half of that same Constitution.

But they are mistaken, because true moderns have always surreptitiously multiplied intermediaries in order to try to conceptualize the massive expansion of hybrids as well as their purification. The sciences have always been as intimately linked to communities as Boyle's pump or Hobbes's Leviathan. *It is the double contradiction that is modern, the contradiction between the two constitutional guarantees of Nature and Society on the one hand, and between the practice of purification and the practice of mediation on the other.* By believing in the total separation of the three terms, by really believing that scientists are extraterrestrials, that matter is immaterial, that technology is ahuman, that politics is pure simulacrum the postmoderns in fact finish off modernism, by definitively taking away the mainspring that had been the source of its tension.

There is only one positive thing to be said about the postmoderns: after them, there is nothing. Far from being the last word, they mark the end of ends – that is, the end of ways of ending and of moving on that led to the succession, at an ever more vertiginous rate, of ever more radical and revolutionary critiques. How could we go further in the absence of tension between Nature and Society, or in the separation between the work of hybridization and that of purification? Will we have to imagine some super-hyper-incommensurability? The 'postmods' are the end of history, and the most amusing part is that they really believe it. And to make quite clear that they are not naive, they claim to be delighted with that end! 'You have nothing to expect from us,' Baudrillard and Lyotard delight in saying. No; indeed. But it is no more in their power to end history than it is not to be naive. They are simply stuck in the impasse of all avant-gardes that have no more troops behind them. Let them sleep till the end of the millennium, as Baudrillard advocates, and let us move on to other things. Or rather, let us retrace our steps. Let us stop moving on.

3.5 Semiotic Turns

While the modernizing philosophies were doing the splits between the two poles of the Constitution in order to absorb the proliferation of quasi-objects, another strategy was being put in place to seize the middle ground, whose dimensions were continuing to expand. Instead of concentrating on the extremes of the work of purification, this strategy concentrated on one of its mediations, language. Whether they are called 'semiotics', 'semiology' or 'linguistic turns', the object of all these philosophies is to make discourse not a transparent intermediary that would put the human subject in contact with the natural world, but a mediator independent of nature and society alike. This autonomization of the sphere of meaning has occupied the best minds of our time for the past half-century. If they too have led us into an impasse, it is not because they have 'forgotten man', or 'abandoned reference', as the modernist reaction is declaring today, but because they themselves have limited their enterprise to discourse alone.

These philosophies have deemed it impossible to autonomize meaning except by bracketing off, on the one hand, the question of reference to the natural world and, on the other, the identity of speaking and thinking subjects. For them, language still occupies that median space of modern philosophy (for Kant, the meeting point of phenomena); but instead of making it more or less transparent or more or less opaque, more or less faithful or more or less treacherous, it has taken over the entire space. Language has become a law unto itself, a law governing itself and its own world. The 'system of language', the 'play of language', the 'signifier', 'writing', the 'text', 'textuality', 'narratives', 'discourse' – these are some of the terms that designate the Empire of Signs – to expand Barthes's title (Barthes, [1970] 1982). While modernizing philosophers were increasingly reviving the distance that separated objects from subjects by making them incommensurable, philosophies of language, discourse or texts were occupying the middle ground that had been left vacant, thinking themselves far removed from the natures and societies that they had bracketed off (Pavel, 1986).

The greatness of these philosophies was that they developed, protected from the dual tyranny of referents and speaking subjects, the concepts that give the mediators their dignity – mediators that are no longer simple intermediaries or simple vehicles conveying meaning from Nature to Speakers, or vice versa. Texts and language make meaning; they even produce references internal to discourse and to the speakers installed within discourse (Greimas, 1976; Greimas and Courtès, 1982). In order to produce natures and societies they need only themselves, and, by a strange bootstrapping operation they extract their principle of reality from other narrative forms. Given the primacy of the signifier, the signifieds bustle about in the vicinity without retaining any special privilege. The text becomes primary; what it expresses or conveys is secondary. Speaking subjects are transformed into so many fictions generated by meaning effects; as for the author, he is no longer anything but the artifact of his own writings (Eco, 1979). The objects being spoken of become reality effects gliding over the surface of the writing. Everything becomes sign and sign system: architecture and cooking, fashion and mythology, politics – even the unconscious itself (Barthes, [1985] 1988).

The great weakness of these philosophies, however, is to render more difficult the connections between an autonomized discourse and what they had provisionally shelved: the referent – on Nature's side – and the speaker – on the side of society/subject. Once again, science studies played their disturbing role. When they applied semiotics to scientific discourse, and not only to literatures of fiction, the autonomization of discourse appeared as an artifice (Bastide, in press). As for rhetoric, it changed its meaning entirely when it had truth and proof to absorb instead of conviction and seduction (Latour, 1987). When we are dealing with science and technology it is hard to imagine for long that we are a text that is writing itself, a discourse that is speaking all by itself, a play of signifiers without signifieds. It is hard to reduce the entire cosmos to a grand narrative, the physics of subatomic particles to a text, subway systems to rhetorical devices, all social structures to discourse. The Empire of Signs lasted no longer than Alexander's, and like Alexander's it was carved up and parcelled out to its generals (Pavel, 1989). Some wanted to render the autonomous system of language more plausible by reestablishing the speaking subject or even the social group, and to that end they went off in search of the old sociology. Others sought to make semiotics less absurd by reestablishing contact with the referent, and they chose the world of science or that of common sense in order to anchor discourse once again. Sociologization, naturalization; the choice is never very broad. Others retained the original impetus of the Empire and set about deconstructing themselves, autonomous glosses on autonomous glosses, to the point of autodissolution.

From this crucial turning point, we have learned that the only way to escape from the parallel traps of naturalization and sociologization consists in granting language its autonomy. Without it, how could we deploy that median space between natures and societies so as to accommodate quasi-objects, quasi-subjects? The various forms of semiotics offer an excellent tool chest for following the mediations of language. But by avoiding the double problem of connections to the referent and connections to the context, they prevent us from following the quasi-objects to the end. These latter, as I have said, are simultaneously real, discursive, and social. They belong to nature, to the collective and to discourse. If one autonomizes discourse by turning nature over to the epistemologists and giving up society to the sociologists, one makes it impossible to stitch these three resources back together.

The postmodern condition has recently sought to juxtapose these three great resources of the modern critique – nature, society and discourse – without even trying to connect them. If they are kept distinct, and if all three are separate from the work of hybridization, the image of the modern world they give is indeed terrifying: a nature and a technology that are absolutely sleek; a society made up solely of false consciousness, simulacra and illusions; a discourse consisting only in meaning effects detached from everything; and this whole world of appearances keeps afloat other disconnected elements of networks that can be combined haphazardly by collage from all places and all times. Enough, indeed, to make one contemplate jumping off a cliff. Here is the cause of the postmoderns' flippant despair, one that has taken over from the angst of their predecessors, masters of the absurd. However, postmoderns would never have reached this degree of derision and dereliction had they not believed – to cap it all – that they had forgotten Being.

3.6 Who Has Forgotten Being?

In the beginning, though, the idea of the difference between Being and beings seemed a fairly good means of harbouring the quasi-objects, a third strategy added to that of the modernizing philosophers and to that of linguistic turns. Quasi-objects do not belong to Nature, or to Society, or to the subject; they do not belong to language, either. By deconstructing metaphysics (that is, the modern Constitution taken in its isolation from the work of hybridization), Martin Heidegger designates the central point where everything holds together, remote from subjects and objects alike. 'What is strange in the thinking of Being is its simplicity. Precisely this keeps us from it' (Heidegger, 1977a). By revolving around this navel, this *omphalos*, the philosopher does assert the existence of an articulation between metaphysical purification and the work of mediation. 'Thinking is on the descent to the poverty of its provisional essence. Thinking gathers language into simple saying. In this way language is the language of Being, as the clouds are the clouds of the sky' (p. 242).

But immediately the philosopher loses this well-intentioned simplicity. Why? Ironically, he himself indicates the reason for this, in an apologue on Heraclitus who used to take shelter in a baker's oven. '*Einai gar kai entautha theous*' – 'here, too, the gods are present,' said Heraclitus to visitors who were astonished to see him warming his poor carcass like an ordinary mortal (Heidegger, 1977b, p. 233). '*Auch hier nämlich wesen Götter an.*' But Heidegger is taken in as much as those naive visitors, since he and his epigones do not expect to find Being except along the Black Forest Holzwege. Being cannot reside in ordinary beings. Everywhere, there is desert. The gods cannot reside in technology – that pure Enframing (Zimmerman, 1990) of being [Ge-Stell], that ineluctable fate [Geschick], that supreme danger [Gefahr]. They are not to be sought in science, either, since science has no other essence but that of technology (Heidegger, 1977b). They are absent from politics, sociology, psychology, anthropology, history – which is the history of Being, and counts its

epochs in millennia. The gods cannot reside in economics – that pure
calculation forever mired in beings and worry. They are not to be found
in philosophy, either, or in ontology, both of which lost sight of their
destiny 2,500 years ago. Thus Heidegger treats the modern world as the
visitors treat Heraclitus: with contempt.

And yet – 'here too the gods are present': in a hydroelectric plant on
the banks of the Rhine, in subatomic particles, in Adidas shoes as well as
in the old wooden clogs hollowed out by hand, in agribusiness as well as
in timeworn landscapes, in shopkeepers' calculations as well as in
Hölderlin's heartrending verse. But why do those philosophers no longer
recognize them? Because they believe what the modern Constitution says
about itself! This paradox should no longer astonish us. The moderns
indeed declare that technology is nothing but pure instrumental mastery,
science pure Enframing and pure Stamping [Das Ge-Stell], that economics is pure calculation, capitalism pure reproduction, the subject pure
consciousness. Purity everywhere! They claim this, but we must be
careful not to take them at their word, since what they are asserting is
only half of the modern world, the work of purification that distils what
the work of hybridization supplies.

Who has forgotten Being? No one, no one ever has, otherwise Nature
would be truly available as a pure 'stock'. Look around you: scientific
objects are circulating simultaneously as subjects objects and discourse.
Networks are full of Being. As for machines, they are laden with subjects
and collectives. How could a being lose its difference, its incompleteness,
its mark, its trace of Being? This is never in anyone's power; otherwise
we should have to imagine that we have truly been modern, we should be
taken in by the upper half of the modern Constitution.

Has someone, however, actually forgotten Being? Yes: anyone who
really thinks that Being has really been forgotten. As Lévi-Strauss says,
'the barbarian is first and foremost the man who believes in barbarism.'
(Lévi-Strauss, [1952] 1987, p. 12). Those who have failed to undertake
empirical studies of sciences, technologies, law, politics, economics,
religion or fiction have lost the traces of Being that are distributed
everywhere among beings. If, scorning empiricism, you opt out of the
exact sciences, then the human sciences, then traditional philosophy,
then the sciences of language, and you hunker down in your forest – then
you will indeed feel a tragic loss. But what is missing is you yourself, not
the world! Heidegger's epigones have converted that glaring weakness
into a strength. 'We don't know anything empirical, but that doesn't
matter, since your world is empty of Being. We are keeping the little
flame of Being safe from everything, and you, who have all the rest, have
nothing.' On the contrary: we have everything, since we have Being, and
beings, and we have never lost track of the difference between Being

and beings. We are carrying out the impossible project undertaken by Heidegger, who believed what the modern Constitution said about itself without understanding that what is at issue there is only half of a larger mechanism which has never abandoned the old anthropological matrix. No one can forget Being, since there has never been a modern world, or, by the same token, metaphysics. We have always remained pre-Socratic, pre-Cartesian, pre-Kantian, pre-Nietzschean. No radical revolution can separate us from these pasts, so there is no need for reactionary counter-revolutions to lead us back to what has never been abandoned. Yes, Heraclitus is a surer guide than Heidegger: '*Einai gar kai entautha theous.*'

3.7 The Beginning of the Past

The proliferation of quasi-objects was thus greeted by three different strategies: first, the ever-increasing separation between the pole of Nature – things-in-themselves – and that of Society or the subject – people-among-themselves; second, the autonomization of language or meaning; finally, the deconstruction of Western metaphysics. Four different resources allow the modern critique to develop these acids: naturalization, sociologization, discursivization, and finally the forgetting of Being. No single one of these resources makes it possible to understand the modern world. If they are put together but kept separate, the situation is still worse, for their results lead only to the ironic despair whose symptom is postmodernism. All these critical resources share the failure to follow both the work of the proliferation of hybrids and the work of purification. In order to exit from the postmoderns' paralysis, it suffices to reutilize all these resources, but they must be pieced together and put to work in shadowing quasi-objects or networks.

But how are we to make these critical resources work together, given that they have emerged only as a result of their disputes with one another? We have to retrace our steps, in order to deploy an intellectual space large enough to accommodate both the tasks of purification and the tasks of mediation – that is, the two halves of the modern world. But how can we retrace our steps? Isn't the modern world marked by the arrow of time? Doesn't it consume the past? Doesn't it break definitively with the past? Doesn't the very cause of the current prostration come precisely from a 'post' modern era that would inevitably succeed the preceding one, which, in a series of catastrophic upheavals, itself succeeded the premodern eras? Hasn't history already ended? By seeking to harbour quasi-objects at the same time as their Constitution, we are obliged to consider the temporal framework of the moderns. Since we refuse to pass 'after' the postmods, we cannot propose to return to a nonmodern world that we have never left, without a modification in the passage of time itself.

We are led from the definition of quasi-objects to that of time, and time too has a modern and a nonmodern dimension, a longitude and a latitude. No one has expressed this better than Charles Péguy in his *Clio*, a stunning meditation on the brewing of history (Péguy, 1961a; see also Latour, 1977). Calendar time may well situate events with respect to a regulated series of dates, but historicity situates the same events with respect to their intensity. This is what the muse of history drolly explains in comparing Victor Hugo's terrible play *Les Burgraves* – an accumulation of time without historicity – to a little phrase of Beaumarchais – a perfect example of historicity without history:

'When I am told that Hatto, the son of Magnus, the Marquis of Verona, the Burgrave of Nollig, is the father of Gorlois, son of Hatto (bastard), Burgrave of Sareck, I learn nothing,' she [Clio] says. 'I do not know them. I shall never know them. But when I am told that Cherubino is dead, *in a swift storming of a fort to which he had not been assigned*, oh, then I really learn something. And I know quite well what I am being told. A secret trembling alerts me to the fact that I have heard.' (p. 276; original emphasis)

The modern passage of time is nothing but a particular form of historicity. Where do we get the idea of time that passes? From the modern Constitution itself. Anthropology is here to remind us: the passage of time can be interpreted in several ways – as a cycle or as decadence, as a fall or as instability, as a return or as a continuous presence. Let us call the interpretation of this passage temporality, in order to distinguish it carefully from time. The moderns have a peculiar propensity for understanding time that passes as if it were really abolishing the past behind it. They all take themselves for Attila, in whose footsteps no grass grows back. They do not feel that they are removed from the Middle Ages by a certain number of centuries, but that they are separated by Copernican revolutions, epistemological breaks, epistemic ruptures so radical that nothing of that past survives in them – nothing of that past ought to survive in them.

'That theory of progress amounts essentially to a theory of savings banks,' says Clio. 'Overall, and universally, it presupposes, it creates an enormous universal savings bank for the entire human community, a huge intellectual savings bank, general and even universal, automatic, for the whole human community, automatic in the sense that humanity would make deposits in it and would never withdraw from it. And in the sense that the contributions would keep on depositing themselves, tirelessly, on their own initiative. Such is the theory of progress. And such are its blueprints. A stepladder.' (Péguy, 1961a, p. 129)

Since everything that passes is eliminated for ever, the moderns indeed sense time as an irreversible arrow, as capitalization, as progress. But since this temporality is imposed upon a temporal regime that works quite differently, the symptoms of discord are multiplied. As Nietzsche observed long ago, the moderns suffer from the illness of historicism. They want to keep everything, date everything, because they think they have definitively broken with their past. The more they accumulate revolutions, the more they save; the more they capitalize, the more they put on display in museums. Maniacal destruction is counterbalanced by an equally maniacal conservation. Historians reconstitute the past, detail by detail, all the more carefully inasmuch as it has been swallowed up for ever. But are we as far removed from our past as we want to think we are? No, because modern temporality does not have much effect on the passage of time. The past remains, therefore, and even returns. Now this resurgence is incomprehensible to the moderns. Thus they treat it as the return of the repressed. They view it as an archaism. 'If we aren't careful,' they think, 'we're going to return to the past; we're going to fall back into the Dark Ages.' Historical reconstitution and archaism are two symptoms of the moderns' incapacity to eliminate what they nevertheless have to eliminate in order to retain the impression that time passes.

If I explain that revolutions attempt to abolish the past but cannot do so, I again run the risk of being taken for a reactionary. This is because for the moderns – as for their antimodern enemies, as well as for their false postmodern enemies – time's arrow is unambiguous: one can go forward, but then one must break with the past; one can choose to go backward, but then one has to break with the modernizing avant-gardes, which have broken radically with their own past. This diktat organized modern thought until the last few years – without, of course, having any effect on the practice of mediation, a practice that has always mixed up epochs, genres, and ideas as heterogeneous as those of the premoderns. If there is one thing we are incapable of carrying out, we now know, it is a revolution, whether it be in science, technology, politics or philosophy. But we are still modern when we interpret this fact as a disappointment, as if archaism had invaded everything, as if there no longer existed any public dump where we could pile up the repressed material behind us. We are still postmodern when we attempt to rise above this disappointment by juxtaposing in a collage elements from all times – elements that are all equally outdated and outmoded.

3.8 The Revolutionary Miracle

What is the connection between the modern form of temporality and the modern Constitution, which tacitly links the two asymmetries of Nature and Society and allows hybrids to proliferate underneath? Why does the modern Constitution oblige us to experience time as a revolution that always has to start over and over again? The answer, once again, has been offered by the daring foray of science studies into history. The social history of science tried to apply the usual tools of cultural history no longer to the soft contingent local human events but to the hard necessary and universal phenomena of Nature. Once again, historians believed that it would be an easy task simply adding a new wing to the castle of history. And, once again, the absorption of sciences forced them to reconsider most of the hidden assumptions of 'normal' history exactly as it had done for the assumptions of sociology, philosophy or anthropology. The modern conception of time, as it is embedded into the discipline of history depends – strangely enough – on a certain conception of science that suppresses the ins and outs of Nature's objects and presents their sudden emergence as if it were miraculous.

Modern time is a succession of inexplicable apparitions attributable to the distinction between the history of sciences or technologies and just plain history. If you suppress Boyle and Hobbes and their disputes, if you eliminate the work of constructing the pump, the domestication of colleagues, the invention of a crossed-out God, the restoration of English Royalty, how are you going to account for Boyle's discovery? The air's spring comes from nowhere. It emerges fully armed. In order to explain what becomes a great mystery, you are going to have to construct an image of time that is adapted to this miraculous emergence of new things that have always already been there, and to human fabrications that no human has ever made. The idea of radical revolution is the only solution the moderns have imagined to explain the emergence of the hybrids that their Constitution simultaneously forbids and allows, and in order to avoid another monster: the notion that things themselves have a history.

There are good reasons for thinking that the idea of political revolution was borrowed from the idea of scientific revolution (Cohen, 1985). We can understand why. How could Lavoisier's chemistry not have been an absolute novelty, since the great scientist eradicated all the traces of his construction and cut all the ties that bound him to his predecessors, whom he relegated to obscurity? That he should have been executed with the same guillotine he had used on his elders, and in the name of the same obscurantist Enlightenment, is a sinister irony of history (Bensaude-Vincent, 1989). The genesis of scientific or technological innovations is so mysterious in the modern Constitution only because the universal transcendence of local and fabricated laws becomes unthinkable, and has to remain so, to avoid a scandal. The history of human beings, for its part, is going to remain contingent, agitated by sound and fury. From now on there will thus be two different histories: one dealing with universal and necessary things that have always been present, lacking any historicity but that of total revolutions or epistemological breaks; the other focusing on the more or less contingent or more or less durable agitation of poor human beings detached from things.

Through this distinction between the contingent and the necessary, the historical and the atemporal, the history of the moderns will be punctuated owing to the emergence of the nonhumans – the Pythagorean theorem, heliocentrism, the laws of gravity, the steam engine, Lavoisier's chemistry, Pasteur's vaccination, the atomic bomb, the computer – and on each occasion time will be reckoned starting from these miraculous beginnings, secularizing each incarnation in the history of transcendent sciences. People are going to distinguish the time 'BC' and 'AC' with respect to computers as they do the years 'before Christ' and 'after Christ'. With the vocal tremors that often accompany declarations on the modern destiny, people even go to the extent of speaking of a 'Judaeo-Christian conception of time', whereas that notion is an anachronism, since neither Jewish mystics nor Christian theologians have had any inclination whatsoever for the modern Constitution. They have constructed their regime of time around Presence (that is, the presence of God), and not around the emergence of the vacuum, or DNA, or microchips, or automated factories. . .

Modern temporality has nothing 'Judaeo-Christian' about it and, fortunately, nothing durable either. It is a projection of the Middle Kingdom on to a line transformed into an arrow by the brutal separation between what has no history but emerges nevertheless in history – the things of nature – and what never leaves history – the labours and passion of humans. *The asymmetry between nature and culture then becomes an asymmetry between past and future. The past was the confusion of things and men; the future is what will no longer confuse them.* Modernization consists in continually exiting from an obscure age that mingled the needs of society with scientific truth, in order to enter into a new age that will finally distinguish clearly what belongs to atemporal nature and what comes from humans, what depends on things and what belongs to signs. Modern temporality arises from a super-position of the difference between past and future with another difference, so much more important, between mediation and purification. The present is outlined by a series of radical breaks, revolutions, which constitute so many irreversible ratchets that prevent us from ever going backward. In itself, this line is as empty as the scansion of a metronome. Yet it is on to this line that the moderns will project the multiplication of quasi-objects and, with the aid of these objects, will trace two series of irreversible advances: one upward, toward progress, and the other downward, toward decadence.

3.9 The End of the Passing Past

The mobilization of the world and of communities on an ever-larger scale multiplies the actors who make up our natures and our societies, but nothing in their mobilization implies an ordered and systematic passage of time. However, thanks to their quite peculiar form of temporality, the moderns will order the proliferation of new actors either as a form of capitalism, an accumulation of conquests, or as an invasion of barbarians, a succession of catastrophes. Progress and decadence are their two great resources, and the two have the same origin. On each of these three lines – calendar time, progress, decadence – it will be possible to locate the antimoderns, who accept modern temporality but reverse its direction. In order to wipe out progress or degeneracy, they want to return toward the past – as if there were a past!

What is the source of the very modern impression that we are living a new time that breaks with the past? Of a liaison, a repetition that in itself has nothing temporal about it (Deleuze, 1968)? The impression of passing irreversibly is generated only when we bind together the cohort of elements that make up our day-to-day universe. It is their systematic cohesion, and the replacement of these elements by others rendered just as coherent in the subsequent period, which gives us the impression of time that passes, of a continuous flow going from the future toward the past – of a stepladder, as Péguy says. Entities have to be made contemporary by moving in step and have to be replaced by other things equally well aligned if time is to become a flow. Modern temporality is the result of a retraining imposed on entities which would pertain to all sorts of times and possess all sorts of ontological statuses without this harsh disciplining.

The vacuum pump in itself is no more modern than it is revolutionary. It associates, combines and redeploys countless actors, some of whom are fresh and novel – the King of England, the Vacuum, the weight of air – but not all of whom can be seen as new. Their cohesiveness is not sufficient to allow a clean break with the past. A whole supplementary work of sorting out, cleaning up and dividing up is required to obtain the impression of a modernization that goes in step with time. If we place

Boyle's discoveries in eternity and they now fall suddenly upon England, if we connect them with those of Galileo and Descartes by linking them in a 'scientific method', and if, finally, we reject Boyle's belief in miracles as archaic, we then get the impression of a radically new modern time. The notion of an irreversible arrow – progress or decadence – stems from an ordering of quasi-objects, whose proliferation the moderns cannot explain. Irreversibility in the course of time is itself due to the transcendence of the sciences and technologies, which indeed escape all comprehension for the moderns, since the two halves of their Constitution are never specified together. It is a classificatory device for dissimulating the inadmissible origin of the natural and social entities from the work of mediation down below. Just as they eliminate the ins and outs of all the hybrids, so the moderns interpret the heterogeneous rearrangements as systematic totalities in which everything would hold together. Modernizing progress is thinkable only on condition that all the elements that are contemporary according to the calendar belong to the same time. For this to be the case, these elements have to form a complete and recognizable cohort. Then, and only then, time forms a continuous and progressive flow, of which the moderns declare themselves the avant-garde and the antimoderns the rearguard while the premoderns are left on the sideline of complete stagnation.

This beautiful order is disturbed once the quasi-objects are seen as mixing up different periods, ontologies or genres. Then a historical period will give the impression of a great hotchpotch. Instead of a fine laminary flow, we will most often get a turbulent flow of whirlpools and rapids. Time becomes reversible instead of irreversible. At first, this does not bother the moderns. They consider everything that does not march in step with progress archaic, irrational or conservative. And as there are antimoderns who are delighted to play the reactionary role that the modern scenario has prepared for them, the great dramas of luminous progress struggling against obscurantism (or the antidrama of the mad revolutionaries against reasonable conservatives) can be deployed, all the same, for the greater pleasure of the spectators. But if the modernizing temporality is to continue to function, the impression of an ordered front of entities sharing the same contemporary time has to remain credible. Thus there must not be too many counter-examples. If they proliferate too much, it becomes impossible to speak of archaism, or of a return of the repressed.

The proliferation of quasi-objects has exploded modern temporality along with its Constitution. The moderns' flight into the future ground to a halt perhaps twenty years ago, perhaps ten, perhaps last year, with the multiplication of exceptions that nobody could situate in the regular flow of time. First, there were the skyscrapers of postmodern architecture –

(architecture is at the origin of this unfortunate expression); then
Khomeini's Islamic revolution, which no one managed to peg as
revolutionary or reactionary. From then on, the exceptions have popped
up without cease. No one can now categorize actors that belong to the
'same time' in a single coherent group. No one knows any longer whether
the reintroduction of the bear in Pyrenees, kolkhozes, aerosols, the Green
Revolution, the anti-smallpox vaccine, Star Wars, the Muslim religion,
partridge hunting, the French Revolution, service industries, labour
unions, cold fusion, Bolshevism, relativity, Slovak nationalism, commer-
cial sailboats, and so on, are outmoded, up to date, futuristic, atemporal,
nonexistent, or permanent. It is this whirlpool in the temporal flow that
the postmoderns have sensed so early and with so much sensitivity in the
two avant-garde movements of fine arts and politics (Hutcheon, 1989).

As always, however, postmodernism is a symptom, not a solution. The
postmoderns retain the modern framework but disperse the elements that
the modernizers grouped together in a well-ordered cluster. The
postmoderns are right about the dispersion; every contemporary
assembly is polytemporal. But they are wrong to retain the framework
and to keep on believing in the requirement of continual novelty that
modernism demanded. By mixing elements of the past together in the
form of collages and citations, the postmoderns recognize to what extent
these citations are truly outdated. Moreover, it is because they are
outmoded that the postmoderns dig them up, in order to shock the
former 'modernist' avant-gardes who no longer know at what altar to
worship. But it is a long way from a provocative quotation extracted out
of a truly finished past to a reprise, repetition or revisiting of a past that
has never disappeared.

3.10 Triage and Multiple Times

Fortunately, nothing obliges us to maintain modern temporality with its
succession of radical revolutions, its antimoderns who return to what
they think is the past, and its double concert of praise and complaint, for
or against continual progress, for or against continual degeneration. We
are not attached for ever to this temporality that allows us to understand
neither our past nor our future, and that forces us to shelve the totality of
the human and nonhuman third worlds. It would be better to say that
modern temporality has stopped passing. Let us not bemoan the fact, for
our real history had only the vaguest of relations with the Procrustean
bed that the modernizers and their enemies imposed on it.

Time is not a general framework but a provisional result of the
connection among entities. Modern discipline has reassembled, hooked

together, systematized the cohort of contemporary elements to hold it together and thus to eliminate those that do not belong to the system. This attempt has failed; it has always failed. There are no longer – there have never been – anything but elements that elude the system, objects whose date and duration are uncertain. It is not only the Bedouins and the !Kung who mix up transistors and traditional behaviours, plastic buckets and animal-skin vessels. What country could not be called 'a land of contrasts'? We have all reached the point of mixing up times. We have all become premodern again. If we can no longer progress in the fashion of the moderns, must we regress in the fashion of the antimoderns? No, we have to pass from one temporality to the other, since a temporality, in itself, has nothing temporal about it. It is a means of connecting entities and filing them away. If we change the classification principle, we get a different temporality on the basis of the same events.

Let us suppose, for example, that we are going to regroup the contemporary elements along a spiral rather than a line. We do have a future and a past, but the future takes the form of a circle expanding in all directions, and the past is not surpassed but revisited, repeated, surrounded, protected, recombined, reinterpreted and reshuffled. Elements that appear remote if we follow the spiral may turn out to be quite nearby if we compare loops. Conversely, elements that are quite contemporary, if we judge by the line, become quite remote if we traverse a spoke. Such a temporality does not oblige us to use the labels 'archaic' or 'advanced', since every cohort of contemporary elements may bring together elements from all times. In such a framework, our actions are recognized at last as polytemporal.

I may use an electric drill, but I also use a hammer. The former is thirty-five years old, the latter hundreds of thousands. Will you see me as a DIY expert 'of contrasts' because I mix up gestures from different times? Would I be an ethnographic curiosity? On the contrary: show me an activity that is homogeneous from the point of view of the modern time. Some of my genes are 500 million years old, others 3 million, others 100,000 years, and my habits range in age from a few days to several thousand years. As Péguy's Clio said, and as Michel Serres repeats, 'we are exchangers and brewers of time' (Serres and Latour, 1992). It is this exchange that defines us, not the calendar or the flow that the moderns had constructed for us. Pile up the burgraves one behind the other, and you will still not have time. Go down sideways to grab hold of the event of Cherubino's death in its intensity, and time will be given unto you.

Are we traditional, then? Not that either. The idea of a stable tradition is an illusion that anthropologists have long since set to rights. The immutable traditions have all budged – the day before yesterday. Most ancestral folklores are like the 'centenary' Scottish kilt, invented out of whole cloth at the beginning of the nineteenth century (Trevor-Roper, 1983), or the Chevaliers du Tastevin of my little town in Burgundy, whose millennial ritual is not fifty years old. 'Peoples without history' were invented by those who thought theirs was radically new (Goody, 1986). In practice, the former innovate constantly; the latter are forced to pass and repass indefinitely through the same rituals of revolutions, epistemological breaks, and quarrel of the Classics against the Moderns. One is not born traditional; one chooses to become traditional by constant innovation. The idea of an identical repetition of the past and that of a radical rupture with any past are two symmetrical results of a single conception of time. We cannot return to the past, to tradition, to repetition, because these great immobile domains are the inverted image of the earth that is no longer promised to us today: progress, permanent revolution, modernization, forward flight.

What are we to do, if we can move neither forward nor backward? Displace our attention. We have never moved either forward or backward. We have always actively sorted out elements belonging to different times. We can still sort. *It is the sorting that makes the times, not the times that make the sorting.* Modernism – like its anti- and post-modern corollaries – was only the provisional result of a selection made by a small number of agents in the name of all. If there are more of us who regain the capacity to do our own sorting of the elements that belong to our time, we will rediscover the freedom of movement that modernism denied us – a freedom that, in fact, we have never really lost. We are not emerging from an obscure past that confused natures and cultures in order to arrive at a future in which the two poles will finally separate cleanly owing to the continual revolution of the present. We have never plunged into a homogeneous and planetary flow arriving either from the future or from the depths of time. Modernization has never occurred. There is no tide, long in rising, that would be flowing again today. There has never been such a tide. We can go on to other things – that is, return to the multiple entities that have always passed in a different way.

3.11 A Copernican Counter-revolution

If we had been able to keep the human multitudes and the nonhuman environment repressed behind us longer, we would probably have been able to continue to believe that modern times were really passing while eliminating everything in their path. But the repressed has returned. The human masses are here again, in the East as well as in the South, and the infinite variety of nonhuman masses have arrived from Everywhere. They can no longer be exploited. They can no longer be surpassed, because nothing surpasses them any longer. There is nothing greater than the nature surrounding us; Eastern peoples can no longer be reduced to their proletarian avant-gardes; as for the Third World masses, nothing will circumscribe them. How can we absorb them? The moderns raise the question in anguish. How can they all be modernized? We might have done it; we thought we could do it; we can no longer believe it possible. Like a great ocean liner that slows down and then comes to a standstill in the Sargasso Sea, the moderns' time has finally been suspended. But time has nothing to do with it. The connections among beings alone make time. It was the systematic connection of entities in a coherent whole that constituted the flow of modern time. Now that this laminary flow has become turbulent, we can give up analyses of the empty framework of temporality and return to passing time – that is, to beings and their relationships, to the networks that construct irreversibility and reversibility.

But how can the principle for classifying entities be changed? How can the illegitimate multitudes be given a representation, a lineage, a civil status? How can this *terra incognita* that is nevertheless so familiar to us be explored? How can we go from the world of objects or that of subjects to what I have called quasi-objects or quasi-subjects? How can we move from transcendent/immanent Nature to a nature that is still just as real, but extracted from the scientific laboratory and then transformed into external reality? How can we shift from immanent/transcendent Society toward collectives of humans and nonhumans? How can we go from the transcendent/immanent crossed-out God to the God of origins who should perhaps be called the God below? How are we to gain access to networks, those beings whose topology is so odd and whose ontology is even more unusual, beings that possess both the capacity to connect and the capacity to divide – that is, the capacity to produce both time and space? How are we to conceptualize the Middle Kingdom? As I have said, we have to trace both the modern dimension and the nonmodern dimension, we have to deploy the latitude and longitude that will allow us to draw maps adapted both to the work of mediation and to the work of purification.

The moderns knew perfectly well how to conceive of this Kingdom. They did not make quasi-objects disappear by eradication and denial, as if they wanted to simply repress them. On the contrary, they recognized their existence but emptied it of any relevance by turning full-blown mediators into mere intermediaries. An intermediary – although recognized as necessary – simply transports, transfers, transmits energy from one of the poles of the Constitution. It is void in itself and can only be less faithful or more or less opaque. A mediator, however, is an original event and creates what it translates as well as the entities between which it plays the mediating role. If we simply restore this mediating role to all the agents, exactly the same world composed of exactly the same entities cease being modern and becomes what it has never ceased to be – that is, nonmodern. How did the modern manage to specify and cancel out the work of mediation both at once? *By conceiving every hybrid as a mixture of two pure forms.* The modern explanations consisted in splitting the mixtures apart in order to extract from them what came from the subject (or the social) and what came from the object. Next they multiplied the intermediaries in order to reconstruct the unity they had broken and wanted none the less to retrieve through blends of pure forms. So these operations of analysis and synthesis always had three aspects: a preliminary purification, a divided separation, and a progressive reblend-ing. The critical explanation always began from the poles and headed toward the middle, which was first the separation point and then the conjunction point for opposing resources – the place of phenomena in Kant's great narrative. In this way the middle was simultaneously maintained and abolished, recognized and denied, specified and silenced. This is why I can say without contradicting myself that no one has ever been modern, and that we have to stop being so. The necessity of multiplying intermediaries to reconstruct the lost unity has always been recognized – thus no one except the postmods really believes in the two extreme poles of Nature and Society radically distinct from free-floating and disconnected networks – but as long as those intermediaries were seen as mixtures made of pure forms, the belief in the existence of a modern world was inescapable. The whole difference hinges on the apparently small nuance between mediators and intermediaries (Hennion, 1991).

If we seek to deploy the Middle Kingdom for itself, we are obliged to invert the general form of the explanations. The point of separation – and conjunction – becomes the point of departure. The explanations no longer proceed from pure forms toward phenomena, but from the centre toward the extremes. The latter are no longer reality's point of attachment, but so many provisional and partial results. The layering of intermediaries is replaced by chains of mediators, according to the model proposed by Antoine Hennion. Instead of denying the existence of hybrids – and reconstructing them awkwardly under the name of intermediaries – this explanatory model allows us instead to integrate the work of purification as a particular case of mediation. The only difference between the modern and nonmodern conception is therefore breached, since purification is considered as a useful work requiring instruments, institutions and know-how whereas in the modern paradigm there was no explicit function and no apparent necessity in the work of mediation.

Kant's Copernican Revolution, as we have seen, offers a perfected model for modernizing explanations, by making the object revolve around a new focus [*foyer*] and multiplying the intermediaries to cancel out distance between the two poles little by little. But nothing obliges us to take that revolution as a decisive event that sets us for ever on the sure path of science, morality and theology. This reversal may be likened to the French Revolution with which it is linked: they are excellent tools for making time irreversible, but they are not irreversible in themselves. I call this reversed reversal – or rather this shift of the extremes centreward and downward, a movement that makes both object and subject revolve around the practice of quasi-objects and mediators – a Copernican counter-revolution. We do not need to attach our explanations to the two pure forms known as the Object or Subject/Society, because these are, on the contrary, partial and purified results of the central practice that is our sole concern. The explanations we seek will indeed obtain Nature and Society, but only as a final outcome, not as a beginning. Nature does revolve, but not around the Subject/Society. It revolves around the collective that produces things and people. The Subject does revolve, but not around Nature. It revolves around the collective out of which people and things are generated. At last the Middle Kingdom is represented. Natures and societies are its satellites.

3.12 From Intermediaries to Mediators

As soon as we bring about the Copernican counter-revolution and place the quasi-object in a position below and equidistant from the former things-in-themselves and the former humans-among-themselves, when we return to our usual practice, we notice that there is no longer any reason to limit the ontological varieties that matter to two (or three, counting the crossed-out God).

Is the vacuum pump that has served as our example hitherto a new ontological variety in its own right? We cannot use asymmetrical historians to answer this question, since they will be unable to locate the common ontological problem. Some will be historians only of seventeenth-century England, and will have no interest whatsoever in the pump except to make it emerge miraculously from the Heaven of Ideas to establish their chronology. On the other side, scientists and epistemologists will describe the physics of the vacuum without paying the slightest attention to England, or even to Boyle. Let us set aside these asymmetrical tasks, one of which ignores nonhumans and the other

humans, and suppose symmetrical historians who will compare the two
balance sheets when using mediators or intermediaries.

In the modern world of the Copernican Revolution there will be no
new entities, since we are supposed to split it in two dividing its
originality among the two poles: a first part would head toward the right
and would become 'Laws of Nature'; a second part would move leftward
and become 'seventeenth-century English society'; while we would mark
the place of the phenomenon, this still empty place where the two poles
have to be stitched back together. Then, by multiplying intermediaries,
we were supposed to take what we had just separated and bring them
closer together. We were to say that the laboratory pump 'reveals' or
'represents' or 'materializes' or 'allows us to grasp' the Laws of Nature.
We were to say, similarly, that the wealthy English gentlemen's
'representations' made it possible to 'interpret' air pressure and to
'accept' the existence of a vacuum. By moving closer to the point of
separation and conjunction, we were to pass from the global context to
the local context, and we were to show how Boyle's gestures and the
pressure of the Royal Society allowed them to understand the defects of
the pump, its leaks and its aberrations. By multiplying intermediary
terms, we were to have ended up reconnecting the two parts that were at
first infinitely distant from Nature and from the Social.

According to such an explanation, nothing essential has happened. To
explain our air pump, we simply plunged a hand alternately either into
the urn that contains for all eternity the beings of Nature, or into the one
that contains the sempiternal mainsprings of the social world. Nature has
always been unchanging. Society always comprises the same resources,
the same interests, the same passions. In the modern perspective, Nature
and Society allow explanation because they themselves do not have to be
explained. Intermediaries exist, of course, and their role is precisely to
establish the link between the two, but they establish links only because
they themselves lack any ontological status. They merely transport,
convey, transfer the power of the only two beings that are real, Nature
and Society. To be sure, they may do a bad job of the transporting; they
may be unfaithful, or obtuse. But their lack of faithfulness does not give
them any importance in their own right, since that is what proves, on the
contrary, their intermediary status. Their competence is not their own. At
worst, they are brutes or slaves; at best, they are loyal servants.

If we bring about the Copernican counter-revolution we are then
obliged to take the work of the intermediaries much more seriously, since
it is no longer their task to transmit the power of Nature and that of
Society, and since they nevertheless all produce the same reality effects. If
we now enumerate the entities endowed with autonomous status, we find
far more than two or three. There are dozens. Does nature abhor a

vacuum or not? Is there a real vacuum in the pump, or could some subtle ether have slipped in? How are the Royal Society's witnesses going to account for the leaks in the air pump? How is the King of England going to consent to let people go back to talking about the properties of matter and reestablishing private cliques just when the question of absolute power is finally about to be resolved? Is the authenticity of miracles supported by the mechanization of matter or not? Is Boyle going to become a respected experimenter if he devotes himself to pursuing these vulgar experimental tasks and abandons the deductive explanation, the only one worthy of a scholar? All these questions are no longer caught between Nature and Society, since they all redefine what Nature may be and what Society is. Nature and Society are no longer explanatory terms but rather something that requires a conjoined explanation. Around the work of the air pump we witness the formation of a new Boyle, a new Nature, a new theology of miracles, a new scholarly sociability, a new Society that will henceforth include the vacuum, scholars, and the laboratory. History does something. Each entity is an event.

We shall no longer explain the innovation of the air pump by reaching alternately into the two urns of Nature and Society. On the contrary, we will refill these urns, or at least profoundly modify their contents. Nature will emerge altered from Boyle's laboratory, and so will English society; but Boyle and Hobbes will also change in the same degree. Such metamorphoses are incomprehensible if only two beings, Nature and Society, have existed from time immemorial, or if the first remains eternal while the second alone is stirred up by history. These metamorphoses become explicable, on the contrary, if we redistribute essence to all the entities that make up this history. But then they stop being simple, more or less faithful, intermediaries. They become mediators – that is, actors endowed with the capacity to translate what they transport, to redefine it, redeploy it, and also to betray it. The serfs have become free citizens once more.

By offering to all the mediators the being that was previously captive in Nature and in Society, the passage of time becomes more comprehensible again. In the world of the Copernican Revolution, in which everything had to be contained between the poles of Nature and Society, history did not really count. Nature was merely discovered, or Society was deployed, or one was applied to the other. Phenomena were nothing but the encounter of already-present elements. There was indeed a contingent history, but for humans alone, detached from the necessity of natural things. As soon as we start from the middle, as soon as we invert the arrows of the explanation, as soon as we take the essence accumulated at the two extremes and redistribute it to the whole set of intermediaries, as soon as we elevate the latter to the status of full-fledged mediators, then history in fact becomes possible. Time is there not for naught, but for real. Something does in fact happen to Boyle, to the air's spring, to the vacuum, to the air pump, to the King, to Hobbes. They all come out changed. All the essences become events, the air's spring by the same token as the death of Cherubino. History is no longer simply the history of people, it becomes the history of natural things as well.

3.13 Accusation, Causation

This Copernican counter-revolution amounts to modifying the place of the object to remove it from things-in-themselves and bring it to the community, but without bringing it closer to society. Michel Serres's work is no less important than Shapin and Schaffer's or Hennion's for achieving this displacement or descent. As Serres writes in one of his best books: 'We want to describe the emergence of the object, not only of tools or beautiful statues, but of things in general, ontologically speaking. How does the object come to what is human?' (Serres, 1987). But his problem is that he 'can't find anything in books that recounts the primitive experience during which the object as such constituted the human subject, because books are written to entomb this very experience, to block all access to it, and because the noise of discourse drowns out what happened in that utter silence' (p. 216).

We possess hundreds of myths describing the way subjects (or the collective, or intersubjectivity, or epistemes) construct the object – Kant's Copernican Revolution is only one in a long line of examples. Yet we have nothing that recounts the other aspect of the story: how objects construct the subject. Shapin and Schaffer have access to thousands of archival pages on Boyle's ideas, and Hobbes's, but nothing about the tacit practice of the air pump or on the dexterity it required. The witnesses to this second half of history are constituted not by texts or languages but by silent, brute remainders such as pumps, stones and statues. Even though Serres's archaeology is situated several levels below that of the air pump, he encounters the same silence.

The people of Israel chant psalms before the dismantled Wailing Wall: of the temple, not one stone remains standing on another. What did the wise Thaleäes see, do, and think, by the Egyptian pyramids, in a time as remote for us as the time of Cheops was for him? Why did he invent geometry by this pile of stones? All Islam dreams of traveling to Mecca where, in the Kaaba, the Black Stone is preserved. Modern science was born, in the Renaissance, from the study of falling bodies: stones fall to the ground. Why did Jesus establish the Christian Church on a man called Peter? I am deliberately mixing religion with science in these examples of inauguration. (Serres, 1987, p. 213)

Why should we take seriously such a hasty generalization of all these petrifications, one that mixes up a black religious stone with Galileo's falling bodies? For the same reason that I have taken Shapin and Schaffer's work seriously, 'deliberately mixing religion and sciences in these examples of inauguration' of modern science and politics. They had ballasted epistemology with an unknown new actor, a leaky, pieced-together, handmade air pump. Serres ballasts epistemology with an unknown new actor, silent things. They all do it for the same anthropological reason: science and religion are linked by a profound reinterpretation of what it means to accuse and put to the test. For Boyle as for Serres, science is a branch of the judiciary:

> In all the languages of Europe, north and south alike, the word 'thing', whatever its form, has as its root or origin the word 'cause', taken from the realm of law, politics, or criticism generally speaking. As if objects themselves existed only according to the debates of an assembly or after a decision issued by a jury. Language wants the world to stem from language alone. At least this is what it says . . . (Serres, 1987, p. 111)

> Thus in Latin the word for 'thing' is *res*, from which we get reality, the object of judicial procedure or the cause itself, so that, for the Ancients, the accused bore the name *reus* because the magistrates were suing him. As if the only human reality came from tribunals alone. (p. 307)

> Here we shall see the miracle and find the solution to the ultimate enigma. The word 'cause' designates the root or origin of the word 'thing': *causa, cosa, chose*, or *Ding*. . . . The tribunal stages the very identity of cause and thing, of word and object, or the passage of one to the other by substitution. A thing emerges there. (p. 294)

Thus with three citations Serres generalizes the results that Shapin and Schaffer brought together with so much difficulty: causes, stones and facts never occupy the position of the thing-in-itself. Boyle wondered how to put an end to civil wars. By compelling matter to be inert, by asking God not to be directly present, by constructing a new closed space in a container where the existence of the vacuum would become manifest, by renouncing the condemnation of witnesses for their opinions. No *ad hominem* accusation will prevail any longer, Boyle said; no human witness will be believed; only nonhuman indicators and instruments observed by gentlemen will be considered trustworthy. The stubborn accumulation of matters of fact will establish the foundations of the pacified collective. This invention of facts is not, however, a discovery of the things that are out there; it is an anthropological creation that redistributes God, will, love, hatred and justice. Serres makes precisely the same point. We have no idea of the aspect things would have outside the tribunal, beyond our civil wars, and outside our trials and our courtrooms. Without accusation we have no causes to plead, and we cannot assign causes to phenomena. This anthropological situation is not limited to our prescientific past, since it belongs more to our scientific present.

Thus we live in a society that is modern not because, unlike all the others, it is finally liberating itself from the hell of social relationships, from the obscurantism of religion, from the tyranny of politics, but because, just like all the others, it is redistributing the accusations that replace a cause – judiciary, collective, social – by a cause – scientific, nonsocial, matter-of-factual. Nowhere can one observe an object and a subject, one society that would be primitive and another that would be modern. Series of substitutions, displacements and translations mobilize peoples and things on an ever-increasing scale.

> I imagine, at the origin, a rapid whirlwind in which the transcendental constitution of the object by the subject would be nourished, as in return, by the symmetrical constitution of the subject by the object, in crushing semicycles that are endlessly begun anew, returning to the origin. . . . There exists a transcendental objective, a constitutive condition of the subject through the appearance of the object as object in general. Of the inverse or symmetrical condition on the whirling cycle, we have testimony, traces or narratives, written in the labile languages. . . . But of the direct constitutive condition on the basis of the object we have witnesses that are tangible, visible, concrete, formidable, tacit. However far back we go in talkative history or silent prehistory, they are still there. (Serres, 1987, p. 209)

Serres, in his so-unmodern work, recounts a pragmatogony that is as fabulous as Hesiod's old cosmogony, or Hegel's. However, Serres proceeds not by metamorphosis or dialectics, but by substitutions. The new sciences that deviate, transform, knead the collective into things that no one has made, are simply latecomers in that long mythology of substitutions. Those who follow networks, or study the sciences, are only documenting the *nth* loop in the spiral whose fabulous beginning Serres sketches for us. Contemporary science is a way of prolonging what we have already done. Hobbes constructs a political body on the basis of animated naked bodies: he finds himself with the gigantic artificial prosthesis of the Leviathan. Boyle concentrates all the dissension of civil wars on an air pump: he finds himself with the facts. Each loop in the spiral defines a new collective and a new objectivity. The collective in permanent renewal that is organized around things in permanent renewal has never stopped evolving. We have never left the anthropological matrix – we are still in the Dark Ages or, if you prefer, we are still in the world's infancy.

3.14 Variable Ontologies

As soon as we grant historicity to all the actors so that we can accommodate the proliferation of quasi-objects, Nature and Society have no more existence than West and East. They become convenient and relative reference points that moderns use to differentiate intermediaries, some of which are called 'natural' and others 'social', while still others are termed 'purely natural' and others 'purely social', and yet others are considered 'not only' natural 'but also' a little bit social. The analysts who head left will be called realists, while those who head right will be called constructivists (Latour, 1992b; Pickering, 1992). Those who want to maintain a position plumb in the middle will invent countless combinations in order to mix Nature with Society (or subjects), alternating the 'symbolic dimension' of things with the 'natural dimension' of societies. Others, more imperialistic or more one-sided, will try to naturalize Society by integrating it into Nature (Hull, 1988), or to socialize Nature by getting it digested by Society (Bloor, [1976] 1991) (or by the Subject, which is more difficult).

Still, these reference points and discussions remain one-dimensional. To classify the entire set of entities along the single line that runs from Nature to Society would amount to drawing cartographic maps on the basis of longitude alone, thereby reducing them to a single line! The second dimension makes it possible to give every latitude to the entities and to deploy the map that registers, as I have said, both the modern Constitution and its practice. How will we define this equivalent of North and South? Mixing my metaphors, I would say that it has to be defined *as a gradient that registers variations in the stability of entities from event to essence*. We still know nothing at all about the air pump when we say that it is the representation of the Laws of Nature, or the representation of English Society, or the product of the two opposite constraints of Nature and Society. We still need to be told whether what is at stake is the air pump as a seventeenth-century event or the air pump as a stabilized essence of the eighteenth century or the twentieth century. The degree of stabilization – the latitude – is as important as the position on the line that runs from the natural to the social – the longitude (see Cussins, 1992, for another and more precise mapping device).

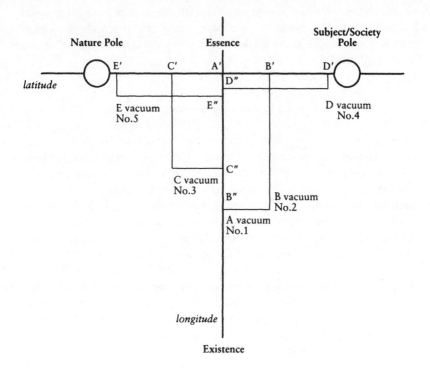

Figure 3.4 The modern Constitution and its practice

The ontology of mediators thus has a variable geometry. What Sartre said of humans – that their existence precedes their essence – has to be said of all the actants: of the air's spring as well as society, of matter as well as consciousness. We do not have to choose between vacuum no. 5, a reality of external nature whose essence does not depend on any human, and vacuum no. 4, a representation that Western thinkers have taken centuries to define. Or rather, we shall be able to choose between the two only once they are stabilized. About the very unstable vacuum no. 1, in Boyle's laboratory, we cannot say whether it is natural or social, only that it emerges artificially in the laboratory. Vacuum no. 2 may be an artifact made by human hands, unless it is transmuted into vacuum no. 3, which begins to become a reality that eludes humans. What is a vacuum, then? None of these positions. The essence of the vacuum is the trajectory that links them all. In other words, the air's spring has a history. Each of the actants possesses a unique signature in the space deployed in this way. In order to trace them, we do not have to form any hypotheses about the essence of Nature or the essence of Society.

Superpose all the signatures and you will have the shapes of what the moderns wrongly call, in order to summarize and purify, 'Nature' and 'Society'.

But if we project all these trajectories on to the single line that connects the former 'Nature' pole with the former 'Society/Subject' pole, everything becomes hopelessly confused. All the points (A, B, C, D, E) will be projected along the single latitude (A', B', C', D', E'), with the central point A localized at the site of the former phenomena – precisely where, in the modern scenario, nothing is supposed to happen, since it is nothing but the meeting point of the two extremes of Nature and Society in which resides the whole of reality. With this single line, realists and constructivists will be able to quarrel over the interpretation of the vacuum for centuries: the former will declare that no one has fabricated this real fact; the latter that our hands alone fashioned this social fact; the advocates of the middle ground will waver between the two senses of the word 'fact', using – for better or for worse – the formula 'not only. . . but also. . .'. This is because the fabrication is below the line, in the work of mediation, visible only if we also take into account the degree of stabilization (B″, C″, D″, E″).

The great masses of Nature and Society can be compared to the cooled-down continents of plate tectonics. If we want to understand their movement, we have to go down into those searing rifts where the magma erupts and on the basis of this eruption are produced – much later and much farther off, by cooling and progressive stacking – the two continental plates on which our feet are firmly planted. Like the geophysicians, we too have to go down and approach the places where the mixtures are made that will become – but only much later – aspects of Nature or of the Social. Is it too much to ask of our discussions that from now on we should spell out the latitude of the entities we are talking about as well as their longitude, and that we should view essences as events and trajectories?

We now have a better understanding of the paradox of the moderns. By using both the work of mediation and the work of purification, but never representing the two together, they were playing simultaneously on the transcendence and the immanence of the two entities, Nature and Society. That gave them four contradictory resources which allowed them an unusual freedom of movement. Now if we draw the map of the ontological varieties, we note that there are not four regions but three. The double transcendence of Nature on one side and Society on the other corresponds to one single set of stabilized essences. For each state of Society there exists a corresponding state of Nature. Nature and Society are not two opposite transcendences but one and the same growing out of the work of mediation. On the other hand, the immanence of naturing-natures and collectives corresponds to a single region: that of the instability of events, that of the work of mediation. The modern Constitution is therefore correct: there is indeed an abyss between Nature and Society, but this abyss is only a delayed result of stabilization. The only abyss that counts separates the work of mediation from the constitutional formatting, but this abyss becomes – owing to the very proliferation of hybrids – a continuous gradient that we are able to traverse as soon as we become once again what we have never stopped being: nonmoderns. If we add to the official, stable version of the Constitution its unofficial, 'hot' or unstable version, the middle is what fills up, on the contrary, and the extremities are emptied. We understand why the nonmoderns are not the successors to the moderns. The former only make official the latter's denied practice. At the price of a little counter-revolution, we finally understand retrospectively what we had always done.

3.15 Connecting the Four Modern Repertoires

By setting up the two dimensions, modern and nonmodern, by operating this Copernican counter-revolution, by making object and subject both slide centreward and downward, perhaps we shall now be able to capitalize on the best resources of the modern critique. The moderns have developed four different repertoires, which they see as incompatible, to accommodate the proliferation of quasi-objects. The first deals with the external reality of a nature of which we are not masters, which exists outside ourselves and has neither our passions nor our desires, even though we are capable of mobilizing and constructing it. The second deals with the social bond, with what attaches human beings to one another, with the passions and desires that move us, with the personified forces that structure society – a society that surpasses us all, even though it is of our own making. The third deals with signification and meaning, with the actants that make up the stories we tell ourselves, with the ordeals they undergo, with the adventures they live through, with the tropes and genres that organize them, with the great narratives that dominate us infinitely, even though they are at the same time merely texts and discourses. The fourth, finally, speaks of Being and deconstructs what we invariably forget when we concern ourselves with beings alone, even though the presence of Being is distributed among beings, is coextensive with their very existence, their very historicity.

These resources are incompatible only in the official version of the Constitution. In practice, we have trouble telling the four apart. We shamelessly confuse our desires with natural entities – that is, with the socially constructed sciences which in turn very much look like discourses and trace our society. As soon as we are on the trail of some quasi-object, it appears to us sometimes as a thing, sometimes as a narrative, sometimes as a social bond, without ever being reduced to a mere being. Our vacuum pump traces the spring of air, but it also sketches in seventeenth-century society and likewise defines a new literary genre, that of the account of a laboratory experiment. In following the pump, do we have to pretend that everything is rhetorical, or that everything is natural, or that everything is socially constructed, or that everything is stamped and stocked? Do we have to suppose that the same pump is in its essence sometimes an object, sometimes a social bond, and sometimes discourse? Or that it is a bit of each? That sometimes it is a mere being, and sometimes it is marked by the ontological difference between Being and beings? And what if it were we ourselves, the moderns, who artificially divided a unique trajectory, which would be at first neither object, nor subject, nor meaning effect, nor pure being? What if the separation of the four repertoires were applied only to stabilized and later stages?

Nothing proves that these resources remain incompatible when we move from essences to events, from purification to mediation, from the modern dimension to the nonmodern dimension, from revolution to the Copernican counter-revolution. Of quasi-objects, quasi-subjects, we shall simply say that they trace networks. They are real, quite real, and we humans have not made them. But they are collective because they attach us to one another, because they circulate in our hands and define our social bond by their very circulation. They are discursive, however; they are narrated, historical, passionate, and peopled with actants of autonomous forms. They are unstable and hazardous, existential, and never forget Being. This liaison of the four repertoires in the same networks once they are officially represented allows us to construct a dwelling large enough to house the Middle Kingdom, the authentic common home of the nonmodern world as well as its Constitution.

The linkage is impossible as long as we remain truly modern, since Nature, Discourse, Society and Being surpass us infinitely, and because these four sets are defined only by their separation, which maintains our constitutional guarantees. But continuity becomes possible if we add to the guarantees the practice of mediation that the Constitution allows because it denies it. The moderns are quite right to want reality, language, society and being all at once. They are wrong only in believing that these sets are forever contradictory. Instead of always analyzing the trajectory of quasi-objects by separating these resources, can we not write as if they ought to be in continuous connection with one another?

We might well escape from the postmodern prostration itself caused by an overdose of the four critical repertoires.

Are you not fed up at finding yourselves forever locked into language alone, or imprisoned in social representations alone, as so many social scientists would like you to be? We want to gain access to things themselves, not only to their phenomena. The real is not remote; rather, it is accessible in all the objects mobilized throughout the world. Doesn't external reality abound right here among us?

Do you not have more than enough of being continually dominated by a Nature that is transcendent, unknowable, inaccessible, exact, and simply true, peopled with entities that lie dormant like the Sleeping Beauty until the day when scientific Prince Charmings finally discover them? The collectives we live in are more active, more productive, more socialized than the tiresome things-in-themselves led us to expect.

Are you not a little tired of those sociologies constructed around the Social only, which is supposed to hold up solely through the repetition of the words 'power' and 'legitimacy' because sociologists cannot cope either with the contents of objects or with the world of languages that nevertheless construct society? Our collectives are more real, more naturalized, more discursive than the tiresome humans-among-themselves led us to expect.

Are you not fed up with language games, and with the eternal scepticism of the deconstruction of meaning? Discourse is not a world unto itself but a population of actants that mix with things as well as with societies, uphold the former and the latter alike, and hold on to them both. Interest in texts does not distance us from reality, for things too have to be elevated to the dignity of narrative. As for texts, why deny them the grandeur of forming the social bond that holds us together?

Are you not tired of being accused of having forgotten Being, of living in a base world emptied of all its substance, all its sacredness and its art? In order to rediscover these treasures, do we really have to give up the historical, scientific and social world in which we live? To apply oneself to the sciences, to technologies, to markets, to things, does not distance us any more from the difference of Being with beings than from society, politics, or language.

Real as Nature, narrated as Discourse, collective as Society, existential as Being: such are the quasi-objects that the moderns have caused to proliferate. As such it behoves us to pursue them, while we simply become once more what we have never ceased to be: amoderns.

4

☐

RELATIVISM

4.1 How to End the Asymmetry

At the beginning of this essay I proposed anthropology as a model for describing our world, since anthropology alone seemed capable of linking up the strange trajectory of quasi-objects as a whole. I quickly recognized, however, that this model was not readily usable, since it did not apply to science and technology. While ethnographers were quite capable of retracing the links that bound the ethnosciences to the social world, they were unable to do so for the exact sciences. In order to understand why it was so difficult to apply the same freedom of tone to the sociotechnological networks of our Western world, I needed to understand what we meant by modern. If we understand modernity in terms of the official Constitution that has to make a total distinction between humans and nonhumans on the one hand and between purification and mediation on the other, then no anthropology of the modern world is possible. But if we link together in one single picture the work of purification and the work of mediation that gives it meaning, we discover, retrospectively, that we have never been truly modern. As a result, the anthropology that has been stumbling over science and technology up to now could once again become the model for description that I have been seeking. Unable to compare premoderns to moderns, it could compare them both to nonmoderns.

Unfortunately, it is not easy to reutilize anthropology as it stands. Shaped by moderns studying people who were said to be premodern, anthropology has internalized, in its practices, concepts and questions, the impossibility I mentioned above. It rules out studying objects of nature, limiting the extent of its inquiries exclusively to cultures. It thus remains asymmetrical. If anthropology is to become comparative, if it is to be able to go back and forth between moderns and nonmoderns, it must be made symmetrical. To this end, it must become capable of confronting not beliefs that do not touch us directly – we are always critical enough of them – but the true knowledge to which we adhere totally. It must therefore be made capable of studying the sciences by surpassing the limits of the sociology of knowledge and, above all, of epistemology.

The first principle of symmetry upset traditional sociology of knowledge by requiring that error and truth be treated in the same terms (Bloor, [1976] 1991). In the past, the sociology of knowledge, by marshalling a great profusion of social factors, had explained only deviations with respect to the straight and narrow path of reason. Error, beliefs, could be explained socially, but truth remained self-explanatory. It was certainly possible to analyze a belief in flying saucers, but not the knowledge of black holes; we could analyze the illusions of parapsychology, but not the knowledge of psychologists; we could analyze Spencer's errors, but not Darwin's certainties. The same social factors could not be applied equally to both. In this double standard we recognize the split in anthropology between sciences, which were not open to study, and ethnosciences, which were.

The presuppositions of the sociology of knowledge would not have intimidated ethnologists for long, if epistemologists – especially in the French tradition – had not erected as a founding principle this same asymmetry between true and false sciences. Only the latter – the 'outdated' sciences – can be related to the social context. As for the 'sanctioned' sciences, they become scientific only because they tear themselves away from all context, from any traces of contamination by history, from any naive perception, and escape even their own past. Here is the difference, for Bachelard and his disciples, between history and the history of sciences (Bachelard, 1967; Canguilhem, [1968] 1988). History may be symmetrical, but that hardly matters, because it never deals with real science; the history of science, on the other hand, must never be symmetrical, because it deals with science and its utmost duty is to make the epistemological break more complete.

A single example will suffice to show to what lengths the rejection of all symmetrical anthropology can be taken when epistemologists have to treat true sciences differently from false beliefs. When Georges Canguilhem distinguishes scientific ideologies from true sciences, he asserts not only that it is impossible to study Darwin – the scientist – and Diderot – the ideologue – in the same terms, but that it must be impossible to lump them together: 'Distinguishing between ideology and science prevents us from seeing continuities where in fact there are only elements of ideology preserved in a science that has supplanted an earlier ideology. Hence such a distinction prevents us from seeing anticipations of the *Origin of Species* in [Diderot's] *Dream of d'Alembert'* (Canguilhem, [1968] 1988 p. 39). Only what breaks for ever with ideology is scientific. It is difficult indeed to pursue the ins and outs of quasi-objects while following such a principle. Once they have passed into the hands of such epistemologists, they will be pulled out by the roots. Objects alone will remain, excised from the entire network that gave them meaning. But why even mention Diderot or Spencer? Why take an interest in error? Because without it the truth would shine too brightly! 'Recognizing the connections between ideology and science should prevent us from reducing the history of science to a featureless landscape, a map without relief' (p. 39). For such epistemologists, 'Whiggish' history is not a mistake to be overcome but a duty to be carried out with utmost rigour. The history of science should not be confused with history (Bowker and Latour, 1987). The false is what makes the true stand out. What Racine did for the Sun King under the lofty name of historian, Canguilhem does for Darwin under the equally usurped label of historian of science.

The principle of symmetry, on the contrary, reestablished continuity, historicity, and – we may as well say it – elementary justice. David Bloor is Canguilhem's opposite number, just as Serres is Bachelard's. 'The only pure myth is the idea of a science devoid of all myth,' writes the latter as he breaks with epistemology (Serres, 1974). For Serres, as for actual historians of science, Diderot, Darwin, Malthus and Spencer have to be explained according to the same principles and the same causes; if you want to account for the belief in flying saucers, make sure your explanations can be used, symmetrically, for black holes (Lagrange, 1990). If you claim to debunk parapsychology, can you use the same factors for psychology (Collins and Pinch, 1982)? If you analyze Pasteur's successes, do the same terms allow you to account for his failures (Latour, 1988b)?

Above all, the first principle of symmetry proposes a slimming treatment for the explanations of errors offered by social scientists. It had become so easy to account for deviation! Society, beliefs, ideology, symbols, the unconscious, madness – everything was so readily available that explanations were becoming obese. But truths? When we lost our facile recourse to epistemological breaks, we soon realized, we who study the sciences, that most of our explanations were not worth much. Asymmetry organized them all, and simply added insult to injury. Everything changes if the staunch discipline of the principle of symmetry forces us to retain only the causes that could serve both truth and falsehood, belief and knowledge, science and parascience. Those who weighed the winners with one scale and the losers with another, while shouting '*vae victis!*' (woe to the vanquished), like Brennus, made that discrepancy incomprehensible up to now. When the balance of symmetry is reestablished with precision, the discrepancy that allows us to understand why some win and others lose stands out all the more sharply.

4.2 The Principle of Symmetry Generalized

The first principle of symmetry offers the incomparable advantage of doing away with epistemological breaks, with *a priori* separations between 'sanctioned' and 'outdated' sciences, or artificial divisions between sociologists who study knowledge, those who study belief systems, and those who study the sciences. Formerly, when the anthropologist returned from his remote land to discover sciences that had been tidied up by epistemology at home, he could establish no continuity between ethnoscience and scientific knowledge. Thus with good reason he abstained from studying nature, and settled for analyzing cultures. Now when he returns and discovers studies – becoming more numerous by the day – that focus on his own sciences and technologies at home, the abyss is already narrower. He can move without too much difficulty from Trobriand navigators to those of the United States Navy (Hutchins, 1980); from calculators in West Africa to arithmeticians in California (Rogoff and Lave, 1984); from technicians in the Ivory Coast to a Nobel laureate in La Jolla (Latour and Woolgar, [1979] 1986); from sacrifices to the god Baal to the Challenger explosion (Serres, 1987). He is no longer required to limit himself to cultures, since Nature – or, rather, natures – have become similarly accessible to study (Pickering, 1992).

However, the principle of symmetry defined by Bloor leads rapidly to an impasse. If it requires an iron discipline in its explanation, the principle itself is asymmetrical, as the following diagram will make clear. Epistemologists and sociologists of knowledge explained truth through its congruence with natural reality, and falsehood through the constraint of social categories, epistemes or interests. They were asymmetrical. Bloor's principle seeks to explain truth and falsehood alike through the same categories, the same epistemes and the same interests. But what terms does it choose? Those that the sciences of society offer social scientists – that is, Hobbes and his many successors. Thus it is asymmetrical not because it separates ideology and science, as epistemologists do, but because it brackets off Nature and makes the 'Society' pole carry the full weight of explanation. Constructivist where Nature is concerned, it is realistic about Society (Callon and Latour, 1992; Collins and Yearley, 1992).

But Society, as we now know, is no less constructed than Nature, since it is the dual result of one single stabilization process. For each state of

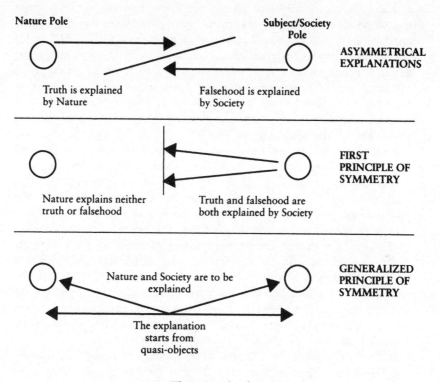

Figure 4.1 The principle of symmetry

Nature there exists a corresponding state of society. If we are to be realist in the one case, we have to be realist in the other; if we are constructivist in one instance, then we have to be constructivist for both. Or rather, as our investigation of the two modern practices has shown, we must be able to understand simultaneously how Nature and Society are immanent – in the work of mediation – and transcendent – after the work of purification. Nature and Society do not offer solid hooks to which we might attach our interpretations (which should be asymmetrical in Canguilhem's sense, or symmetrical in Bloor's), but are what is to be explained. The appearance of explanation that Nature and Society provide comes only in a late phase, when stabilized quasi-objects have become, after cleavage, objects of external reality on the one hand, subjects of Society on the other. Nature and Society are part of the problem, not part of the solution.

If anthropology is to become symmetrical, therefore, it has to do more than take in the first principle of symmetry – which puts a stop to only the most flagrant injustices of epistemology. It has to absorb what Michel

Callon calls the principle of generalized symmetry: the anthropologist has to position himself at the median point where he can follow the attribution of both nonhuman and human properties (Callon, 1986). He is not allowed to use external reality to explain society, or to use power games to account for what shapes external reality. In the same way, he is of course forbidden to alternate natural realism and sociological realism by using 'not only' Nature 'but also' Society, in order to keep the two original asymmetries even while concealing the weaknesses of the one under those of the other (Latour, 1987).

So long as we were modern, it was impossible to occupy this central place from which the symmetry between Nature and Society becomes visible at last, because it did not exist! The only central position recognized by the Constitution, as we have already seen, was the phenomenon, the meeting point where the Nature pole and the Subject pole were applied to one another. Hitherto this point has remained a no-man's-land, a nonplace. Everything changes when, instead of constantly and exclusively alternating between one pole of the modern dimension and the other, we move down along the nonmodern dimension. The unthinkable nonplace becomes the point in the Constitution where the work of mediation emerges. It is far from empty: quasi-objects, quasi-subjects, proliferate in it. No longer unthinkable, it becomes the terrain of all the empirical studies carried out on the networks.

But isn't this place the one that anthropology prepared so painstakingly over the course of a century, the one the ethnologist occupies so effortlessly today when she sets out to study other cultures? Indeed, we can watch her move, without modifying her analytical tools, from meteorology to the kinship system, from the nature of plants to their cultural representation, from political organization to ethnomedicine, from mythic structures to ethnophysics or to hunting techniques. To be sure, the ethnologist draws the courage to deploy this seamless web from her profound conviction that she is dealing merely, and solely, with representations. Nature, for its part, remains unique, external and universal. But if we superpose the two positions – the one that the ethnologist occupies effortlessly in order to study cultures and the one that we have made a great effort to define in order to study our own nature – then comparative anthropology becomes possible, if not easy. It no longer compares cultures, setting aside its own, which through some astonishing privilege possesses a unique access to universal Nature. *It compares natures-cultures.* Are they comparable? Are they similar? Are they the same? We can now, perhaps, solve the insoluble problem of relativism.

4.3 The Import – Export System of the Two Great Divides

'We Westerners are absolutely different from others!' – such is the moderns' victory cry, or protracted lament. The Great Divide between Us – Occidentals – and Them – everyone else, from the China seas to the Yucatan, from the Inuit to the Tasmanian aborigines – has not ceased to obsess us. Whatever they do, Westerners bring history along with them in the hulls of their caravels and their gunboats, in the cylinders of their telescopes and the pistons of their immunizing syringes. They bear this white man's burden sometimes as an exalting challenge, sometimes as a tragedy, but always as a destiny. They do not claim merely that they differ from others as the Sioux differ from the Algonquins, or the Baoules from the Lapps, but that they differ radically, absolutely, to the extent that Westerners can be lined up on one side and all the cultures on the other, since the latter all have in common the fact that they are precisely cultures among others. In Westerners' eyes the West, and the West alone, is not a culture, not merely a culture.

Why does the West see itself this way? Why would the West and only the West not be a culture? In order to understand the Great Divide between Us and Them, we have to go back to that other Great Divide between humans and nonhumans that I defined above. In effect, *the first is the exportation of the second.* We Westerners cannot be one culture among others, since we also mobilize Nature. We do not mobilize an image or a symbolic representation of Nature, the way the other societies do, but Nature as it is, or at least as it is known to the sciences – which remain in the background, unstudied, unstudiable, miraculously conflated with Nature itself. Thus at the heart of the question of relativism we find the question of science. If Westerners had been content with trading and conquering, looting and dominating, they would not distinguish themselves radically from other tradespeople and conquerors. But no, they invented science, an activity totally distinct from conquest and trade, politics and morality.

Even those who have tried, in the name of cultural relativism, to defend the continuity of cultures without ordering them in a progressive series, and without isolating them in their separate prisons (Lévi-Strauss, [1952] 1987), think they can do this only by bringing them as close as possible to the sciences.

'We have had to wait until the middle of this century', writes Lévi-Strauss in *The Savage Mind*, 'for the crossing of long separated paths: that which arrives at the physical world by the detour of communication [the savage mind], and that which, as we have recently come to know, arrives at the world of communication by the detour of the physical [modern science]' (Lévi-Strauss, [1962] 1966, p. 269).

The false antimony between logical and prelogical mentality was sur-
mounted at the same time. The savage mind is as logical in the same sense
and the same fashion as ours, though as our own is only when it is applied
to knowledge of a universe in which it recognizes physical and semantic
properties simultaneously . . . It will be objected that there remains a major
difference between the thought of primitives and our own: Information
Theory is concerned with genuine messages whereas primitives mistake
mere manifestations of physical determinism for messages . . . In treating
the sensible properties of the animal and plant kingdoms as if they were the
elements of a message, and in discovering 'signatures' – and so signs – in
them, men [those with savage minds] have made mistakes of identification:
the meaningful element was not always the one they supposed. But,
without perfected instruments which would have permitted them to place it
where it most often is – namely, at the microscopic level – they already
discerned 'as through a glass darkly' principles of interpretation whose
heuristic value and accordance with reality have been revealed to us only
through very recent inventions: telecommunications, computers and
electron microscopes. (Lévi-Strauss, [1962] 1966, p. 268)

Lévi-Strauss, a generous defence lawyer, imagines no mitigating circumstances other than making his clients look as much like scientists as
possible! If primitive peoples do not differ from us as much as we think, it
is because they anticipate the newest conquests of information theory,
molecular biology and physics, but with inadequate instruments and
'errors of identification'. The very sciences that are used for this
promotion are now off limits. Conceived in the fashion of epistemology,
these sciences remain objective and external, quasi-objects purged of
their networks. Give the primitives a microscope, and they will think
exactly as we do. Is there a better way to finish off those one wants to
save from condemnation? For Lévi-Strauss (as for Canguilhem, Lyotard,
Girard, Derrida, and the majority of French intellectuals), this new
scientific knowledge lies entirely outside culture. It is the transcendence
of science – conflated with Nature – that makes it possible to relativize all
cultures, theirs and ours alike – with the one caveat, of course, that it is
precisely our culture, not theirs, that is constructed through biology,
electronic microscopes and telecommunication networks. . . . The abyss
that was to supposed to be narrowing opens up again.

Somewhere in our societies, and in ours alone, an unheard-of
transcendence has manifested itself: Nature as it is, ahuman, sometimes
inhuman, always extrahuman. Since this event occurred – whether one
situates it in Greek mathematics, Italian physics, German chemistry,
American nuclear engineering or Belgian thermodynamics – there has
been a total asymmetry between the cultures that took Nature into
account and those that took into account only their own culture or the

distorted versions that they might have of matter. Those who invent sciences and discover physical determinisms never deal exclusively with human beings, except by accident. The others have only representations of Nature that are more or less disturbed or coded by the cultural preoccupations of the humans that occupy them fully and fall only by chance – 'as through a glass darkly' – on things as they are.

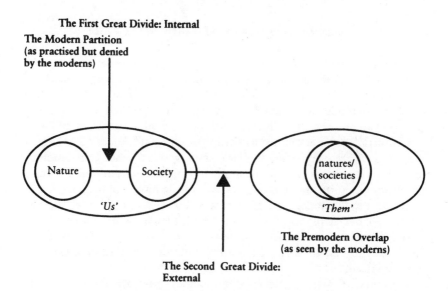

Figure 4.2 The two Great Divides

So the Internal Great Divide accounts for the External Great Divide: we are the only ones who differentiate absolutely between Nature and Culture, between Science and Society, whereas in our eyes all the others – whether they are Chinese or Amerindian, Azande or Barouya – cannot really separate what is knowledge from what is Society, what is sign from what is thing, what comes from Nature as it is from what their cultures require. Whatever they do, however adapted, regulated and functional they may be, they will always remain blinded by this confusion; they are prisoners of the social and of language alike. Whatever we do, however criminal, however imperialistic we may be, we escape from the prison of the social or of language to gain access to things themselves through a providential exit gate, that of scientific knowledge. The internal partition between humans and nonhumans defines a second partition – an external one this time – through which the moderns have set themselves apart from the premoderns. For Them, Nature and Society, signs and things,

are virtually coextensive. For Us they should never be. Even though we
might still recognize in our own societies some fuzzy areas in madness,
children, animals, popular culture and women's bodies (Haraway, 1989),
we believe our duty is to extirpate ourselves from those horrible mixtures
as forcibly as possible by no longer confusing what pertains to mere
social preoccupations and what pertains to the real nature of things.

4.4 Anthropology Comes Home from the Tropics

When anthropology comes home from the tropics in order to rejoin the
anthropology of the modern world that is ready and waiting, it does so at
first with caution, not to say with pusillanimity. At first, it thinks it can
apply its methods only when Westerners mix up signs and things the way
savage thought does. It will therefore look for what most resembles its
traditional terrains as defined by the External Great Divide. To be sure, it
has to sacrifice exoticism, but not at great cost, since anthropology
maintains its critical distance by studying only the margins and fractures
of rationality, or the realms beyond rationality. Popular medicine,
witchcraft in the Bocage (Favret-Saada, 1980), peasant life in the shadow
of nuclear power plants (Zonabend, 1989), the representations ordinary
people have of technical risks (Douglas, 1983) – all these can be excellent
field study topics, because the question of Nature – that is, of science – is
not yet raised.

However, the great repatriation cannot stop there. In fact, by
sacrificing exoticism, the ethnologist loses what constituted the very
originality of her research as opposed to the scattered studies of
sociologists, economists, psychologists or historians. In the tropics, the
anthropologist did not settle for studying the margins of other cultures
(Geertz, 1971). If she remained marginal by vocation and method, and
out of necessity, she nevertheless claimed to be reconstituting the centre
of those cultures: their belief system, their technologies, their ethnosciences, their power plays, their economies – in short, the totality of
their existence (Mauss, [1923] 1967). If she comes back home but limits
herself to studying the marginal aspects of her own culture, she loses all
the hard-won advantages of anthropology. For example Marc Augé
when he resided among the lagoon-dwellers of the Ivory Coast, sought to
understand the entire social phenomenon revealed by sorcery (Augé,
1975). His marginality did not hinder him from grasping the full social
fabric of Alladian culture. But back at home he has limited himself to
studying the most superficial aspects of the metro (Augé, 1986),
interpreting some graffiti on the walls of subway corridors, intimidated
this time by the evidence of his own marginality in the face of Western

economics, technologies and science. A symmetrical Marc Augé would have studied the sociotechnological network of the metro itself: its engineers as well as its drivers, its directors and its clients, the employer-State, the whole shebang – simply doing at home what he had always done elsewhere. Western ethnologists cannot limit themselves to the periphery; otherwise, still asymmetrical, they would show boldness toward others, timidity toward themselves. Back home anthropology need not become the marginal discipline of the margins, picking up the crumbs that fall from the other disciplines' banquet table.

In order to achieve such freedom of movement and tone, however, one has to be able to view the two Great Divides in the same way, and consider them both as one particular definition of our world and its relationships with the others. Now these Divides do not define us any better than they define others; they are no more an instrument of knowledge than is the Constitution alone, or modern temporality alone (see Section 3.7). To become symmetrical, anthropology needs a complete overhaul and intellectual retooling so that it can get around both Divides at once by believing neither in the radical distinction between humans and nonhumans at home, nor in the total overlap of knowledge and society elsewhere.

Let us imagine an ethnologist who goes out to the tropics and takes along with her the Internal Great Divide. In her eyes, the people she studies continually confuse knowledge of the world – which the investigator, as a good scientistic Westerner, possesses as her birthright – and the requirements of social functioning. The tribe that greets her thus has only one vision of the world, only one representation of Nature. To go back to the expression Marcel Mauss and Emile Durkheim made famous, this tribe projects its own social categories on to Nature (Durkheim and Mauss, [1903] 1967; Haudricourt, 1962). When our ethnologist explains to her informers that they must be more careful to separate the world as it is from the social representation they provide for it, they are scandalized or nonplussed. The ethnologist sees in their rage and their misunderstanding the very proof of their premodern obsession. The dualism in which she lives – humans on one side, nonhumans on the other, signs over here, things over there – is intolerable to them. For social reasons, our ethnologist concludes, this culture requires a monist attitude. 'We traffic in ideas; [the savage mind] hoards them up' (Lévi-Strauss, [1962] 1966, p. 267).

But let us suppose now that our ethnologist returns to her homeland and tries to dissolve the Internal Great Divide. And let us suppose that through a series of happy accidents she sets out to analyze one tribe among others – for example, scientific researchers or engineers (Knorr-Cetina, 1992). The situation turns out to be reversed, because now she applies the lessons of monism she thinks she has learned from her earlier experience. Her tribe of scientists claims that in the end they are completely separating their knowledge of the world from the necessities of politics and morality (Traweek, 1988). In the observer's eyes, however, this separation is never very visible, or is itself only the by-product of a much more mixed activity, some tinkering in and out of the laboratory. Her informers claim that they have access to Nature, but the ethnographer sees perfectly well that they have access only to a vision, a representation of Nature that she herself cannot distinguish neatly from politics and social interests (Pickering, 1980). This tribe, like the earlier one, projects its own social categories on to Nature; what is new is that it pretends it has not done so. When the ethnologist explains to her informers that they cannot separate Nature from the social representation they have formed of it, they are scandalized or nonplussed. Our ethnologist sees in their rage and incomprehension the very proof of their modern obsession. The monism in which she now lives – humans are always mixed up with nonhumans – is intolerable to them. For social reasons, our ethnologist concludes, Western scientists require a dualist attitude.

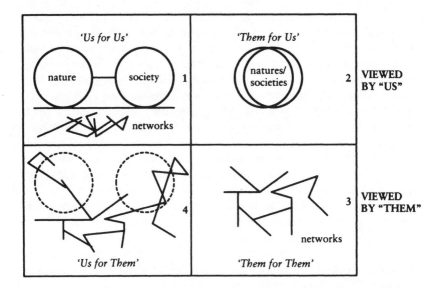

Figure 4.3 Them and Us

However, her double conclusion is incorrect, for she has not really heard what her informers were saying. The goal of anthropology is not to scandalize twice over, or to provoke incomprehension twice in a row: the first time by exporting the Internal Great Divide and imposing dualism on cultures that reject it; the second time by cancelling the External Great Divide and imposing monism on a culture, our own – that rejects it absolutely. Symmetrical anthropology must realize that the two Great Divides do not describe reality – our own as well as that of others – but define the particular way Westerners had of establishing their relations with others as long as they felt modern. 'We', however, do not distinguish between Nature and Society more than 'They' make them overlap. If we take into account the networks that we allow to proliferate beneath the official part of our Constitution they look a lot like the networks in which 'They' say they live. Premoderns are said never to distinguish beween signs and things, but neither do 'We' (Figure 4.3.3 and the bottom of 4.3.1 look very much alike). If, through an acrobatic thought experiment, we could go further and ask 'Them' to try to map on to their own networks our strange obsession with dichotomies and to try to imagine, in their own terms, what it could mean to have a pure Nature and a pure Society they would draw, with extreme difficulty, a provisional map in which Nature and Society would barely escape from the networks (Figure 4.3.4). But what does this picture represent, this picture in which Nature and Culture appear to be redistributed among the networks and to escape from them only fuzzily as if in dotted lines? It is exactly our world as we now see it through nonmodern eyes! It is exactly the picture I have tried to offer from the beginning, in which the upper and lower halves of the Constitution gradually merge. Premoderns are like us. Once they are considered symmetrically, they might offer a better analysis of the Westerners than the modernist anthropology offered of the premoderns! Or, more exactly, we can now drop entirely the 'Us' and 'Them' dichotomy, and even the distinction between moderns and premoderns. We have both always built communities of natures and societies. There is only one, symmetrical, anthropology.

4.5 There Are No Cultures

Let us suppose that anthropology, having come home from the tropics, sets out to retool itself by occupying a triply symmetrical position. It uses the same terms to explain truths and errors (this is the first principle of symmetry); it studies the production of humans and nonhumans simultaneously (this is the principle of generalized symmetry); finally, it refrains from making any *a priori* declarations as to what might distinguish Westerners from Others. To be sure, it loses exoticism, but it gains new fields of study that allow it to analyze the central mechanism of all collectives, including the ones to which Westerners belong. It loses its

exclusive attachment to cultures alone – or to cultural dimensions alone
– but it gains a priceless acquisition, natures. The two positions I have
been staking out since the beginning of this essay – the one the
ethnologist is now occupying effortlessly, and the one the analyst of the
sciences was striving toward with great difficulty – can now be
superimposed. Network analysis extends a hand to anthropology, and
offers it the job that has been ready and waiting.

The question of relativism is already becoming less difficult. If science
as conceived along the epistemologists' lines made the problem insoluble,
it suffices, as is often the case, to change the conception of scientific
practices in order to dispel the artificial difficulties. What reason
complicates, networks explicate. It is the peculiar trait of Westerners that
they have imposed, by their official Constitution, the total separation of
humans and nonhumans – the Internal Great Divide – and have thereby
artificially created the scandal of the others. 'How can one be a Persian?'
How can one not establish a radical difference between universal Nature
and relative culture? But *the very notion of culture is an artifact created
by bracketing Nature off*. Cultures – different or universal – do not exist,
any more than Nature does. There are only natures-cultures, and these
offer the only possible basis for comparison. As soon as we take practices
of mediation as well as practices of purification into account, we discover
that the moderns do not separate humans from nonhumans any more
than the 'others' totally superimpose signs and things.

I can now compare the forms of relativism according to whether they
do or do not take into account the construction of natures as well.
Absolute relativism presupposes cultures that are separate and incommensurable and cannot be ordered in any hierarchy; there is no use
talking about it, since it brackets off Nature. As for cultural relativism,
which is more subtle, Nature comes into play, but in order to exist it does
not presuppose any scientific work, any society, any construction, any
mobilization, any network. It is Nature revisited and corrected by
epistemology, for which scientific practice still remains off camera, *hors
champ*. Within this tradition, the cultures are thus distributed as so many
more or less accurate viewpoints on that unique Nature. Certain societies
see it 'as through a glass darkly', others see it through thick fog, still others
under clear skies. Rationalists will insist on the common aspects of all
these viewpoints; relativists will insist on the irresistible distortion that
social structures impose on all perception. The former will be undone if it
can be shown that cultures do not superimpose their categories; the latter
will lose ground if it can be proved that the categories are superimposed
(Hollis and Lukes, 1982; Wilson, 1970).

In practice, however, as soon as Nature comes into play without being
attached to a particular culture, a third model is always secretly used : a

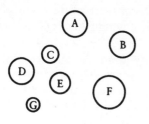

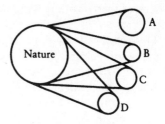

ABSOLUTE RELATIVISM

Culture without hierarchy
and without contacts,
all incommensurable;
Nature is bracketed

CULTURAL RELATIVISM

Nature is present but outside cultures;
cultures all have a more or less precise
point of view toward Nature

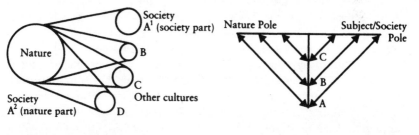

PARTICULAR UNIVERSALISM

One of the cultures (A) has
a privileged access to Nature
which sets it apart from the others

SYMMETRICAL ANTHROPOLOGY

All the collectives similarly constitute
natures and cultures; only the
scale of the mobilization varies

Figure 4.4 Relativism and universalism

type of universalism that I would call 'particular'. One society – and it
is always the Western one – defines the general framework of Nature
with respect to which the others are situated. This is Lévi-Strauss's
solution: he distinguishes Western society, which has a specific interpretation of Nature, from that Nature itself, miraculously known to our
society. The first half of the argument allows for modest relativism (we
are just one interpretation among others), but the second permits the
surreptitious return of arrogant universalism – we remain absolutely
different. In Lévi-Strauss's eyes, however, there is no contradiction
between the two halves, precisely because our Constitution, and it alone,
allows us to distinguish society A^1, made up of humans, from society A^2,
composed of nonhumans but forever removed from the first one! The

contradiction stands out today only in the eyes of symmetrical anthropology. This latter model is the common stock of the other two, whatever the relativists (who never relativize anything but cultures) may say.

The relativists have never been convincing on the subject of the equality of cultures, since they limit their consideration precisely to cultures. And Nature? According to them, it is the same for all, since universal science defines it. In order to get out of this contradiction, they then either have to limit all peoples to a representation of the world by locking them up for ever in the prison of their own societies or, conversely, they have to reduce all scientific results to products of local and contingent social constructions in order to deny science any universality. But to imagine billions of people imprisoned in distorted views of the world since the beginning of time is as difficult as it is to imagine neutrinos and quasars, DNA and universal gravitation, as Texan, British or Burgundian social productions. The two responses are equally absurd, and that is why the great debates over relativism never lead anywhere. *It is as impossible to universalize nature as it is to reduce it to the narrow framework of cultural relativism alone.*

The solution appears along with the dissolution of the artifact of cultures. All natures-cultures are similar in that they simultaneously construct humans, divinities and nonhumans. None of them inhabits a world of signs or symbols arbitrarily imposed on an external Nature known to us alone. None of them – and especially not our own – lives in a world of things. All of them sort out what will bear signs and what will not. If there is one thing we all do, it is surely that we construct both our human collectives and the nonhumans that surround them. In constituting their collectives, some mobilize ancestors, lions, fixed stars, and the coagulated blood of sacrifice; in constructing ours, we mobilize genetics, zoology, cosmology and hæmatology. 'But those are sciences!' the moderns will exclaim, horrified at this confusion. 'They have to escape the representations of society to the greatest possible extent!' Yet the presence of the sciences does not suffice to break the symmetry; such is the discovery of comparative anthropology. From cultural relativism we move on to 'natural' relativism. The first led to absurdities; the second will allow us to fall back on common sense.

4.6 Sizeable Differences

Still, the problem of relativism has not been solved. Only the confusion resulting from the bracketing off of Nature has been provisionally eliminated. We now find ourselves confronting productions of natures-cultures that I am calling collectives – as different, it should be recalled, from the society construed by sociologists – men-among-themselves – as they are from the Nature imagined by epistemologists – things-in-themselves. In the view of comparative anthropology these collectives are all alike, as I have said, in that they distribute both what will later, after stabilization, become elements of Nature and elements of the social world. No one has ever heard of a collective that did not mobilize heaven and earth in its composition, along with bodies and souls, property and law, gods and ancestors, powers and beliefs, beasts and fictional beings. . . . Such is the ancient anthropological matrix, the one we have never abandoned.

But this common matrix defines only the point of departure of comparative anthropology. All collectives are different from one another in the way they divide up beings, in the properties they attribute to them, in the mobilization they consider acceptable. These differences constitute countless small divides, and there is no longer a Great Divide to take one apart from all the others. Among these small divides, there is one that we are now capable of recognizing as such, one that has distinguished the official version of certain segments of certain collectives for three centuries. This is our Constitution, which attributes the role of nonhumans to one set of entities, the role of citizens to another, the function of an arbitrary and powerless God to a third and cuts off the work of mediation from that of purification. In itself this Constitution does not separate us significantly from others, since it is added to the long list of differential traits that define us in the eyes of comparative anthropology. Those traits could be transcribed as a set of entries in the huge data base of anthropology departments – which would then simply have to be rechristened 'Human and Nonhuman Relations Area Files'!

In our distribution of variable-geometry entities, we are as different from the Achuar as they are from the Tapirapè or the Arapesh. No more so, and no less. Such a comparison, however, respects only the conjoined production of one nature-culture, which is only one aspect of collectives. It may satisfy our sense of justice, but in various ways it encounters the same difficulty as absolute relativism, since it immediately abolishes differences by rendering them all equally different. It does not allow us to account for that other aspect of what I have been pursuing since the beginning of this essay – the scope of the mobilization, a scope that issimultaneously the consequence of modernism and the cause of its demise.

This is because the principle of symmetry aims not only at establishing equality – which is only the way to set the scale at zero – but at registering differences – that is, in the final analysis, asymmetries – and at understanding the practical means that allow some collectives to

dominate others. Even though they might be similar in the principle of
their co-production, collectives may differ in size. At the beginning of the
weighing-in process, a nuclear power plant, or a hole in the ozone layer,
or a map of the human genome, or a rubber-tyred metro train, or a
satellite network, or a cluster of galaxies, weighs no more than a wood
fire, or the sky that may fall on our heads, or a genealogy, or a cart, or
spirits visible in the heavens, or a cosmogony. As I said above, this is not
yet enough to break the symmetry. In each case these quasi-objects trace,
with their hesitant trajectories, both forms of nature and forms of
society. When, however, the weighing is complete, the first lot outlines an
entirely different collective from the second. These new differences,
measurable only because the scales have first been calibrated by the
principle of symmetry, have to be recognized as well.

In other words, the differences are sizeable, but they are only of size.
They are important (and the error of cultural relativism is that it ignores
them), but they are not disproportionate (and the error of universalism is
that it sets them up as a Great Divide). The collectives are all similar,
except for their size, like the successive helixes of a single spiral. The fact
that one of the collective needs ancestors and fixed stars while another
one, more eccentric, needs genes and quasars, is explained by the
dimensions of the collective to be held together. A much larger number of
objects requires a much larger number of subjects. A much greater degree
of subjectivity requires a much greater degree of objectivity. If you want
Hobbes and his descendants, you have to take Boyle and his as well. If
you want the Leviathan, you have to have the air pump too. This is the
stance that makes it possible to respect the differences (the dimensions of
the helixes do vary) while at the same time respecting the similarities (all
collectives mix human and nonhuman entities together in the same way).
Relativists, who strive to put all cultures on an equal footing by viewing
all of them as equally arbitrary codings of a natural world whose
production is unexplained, do not succeed in respecting the efforts
collectives make to dominate one another. And universalists on the other
hand, are incapable of understanding the deep fraternity of collectives,
since they are obliged to offer access to Nature to Westerners alone, and
to imprison all others in social categories from which they will escape
only by becoming scientific, modern and Westernized.

Sciences and technologies are remarkable not because they are true or
efficient – they gain these properties in addition, and for reasons entirely
different from those the epistemologists provide (Latour, 1987) – but
because they multiply the nonhumans enrolled in the manufacturing of
collectives and because they make the community that we form with
these beings a more intimate one. The extension of the spiral, the scope of
the enlistments it will bring about, the ever-increasing lengths to which it

goes to recruit these beings, are what characterize the modern sciences, not some epistemological break that would cut them off for ever from their prescientific past. Modern knowledge and power are different not in that they would escape at last the tyranny of the social, but in that *they add many more hybrids in order to recompose the social link and extend its scale*. Not only the air pump but also microbes, electricity, atoms, stars, second-degree equations, automatons and robots, mills and pistons, the unconscious and neurotransmitters. At each turn in the spiral, a new translation of quasi-objects gives new impetus to the redefinition of the social body, of subjects and objects alike. Sciences and technologies, for 'Us', do not reflect society any more than Nature reflects social structures for 'Them'. No one is fiddling with mirrors. It is a matter of constructing collectives themselves on scales that grow larger and larger. There are indeed differences, but they are differences in size. There are no differences in nature – still less in culture.

4.7 Archimedes' *coup d'état*

What explains this new asymmetry which the principle of symmetry, generalized, allows us to detect? The relative size of collectives will be profoundly modified by the enlistment of a particular type of non-humans. To help us understand this variation in size, there is no more striking emblem than an impossible experiment recounted by Plutarch – Michel Authier has called it 'the canon of the savant' (Authier, 1989), and it is as striking as Boyle's air pump:

> Archimedes, who was a kinsman and friend of King Hiero, wrote to him that with any given force it was possible to move any given weight; and emboldened, as we are told, by the strength of his demonstration, he declared that if there were another Earth, and he could go to it, he could move this one. Hiero was astonished and begged him to put his proposition into execution, and show him some great weight moved by a slight force. Archimedes therefore fixed upon a three-masted merchantman of the royal fleet, which had been dragged ashore by the great labours of many men, and after putting on board many passengers and the customary freight, he seated himself at a distance from her, and without any great effort, but quietly setting in motion with his hand a system of compound pulleys, drew her towards him smoothly and evenly, as though she were gliding through the water. Amazed at this, then, and comprehending the power of his art, the King persuaded Archimedes to prepare for him offensive and defensive engines to be used in every kind of siege warfare. (Plutarch, *Marcellus' Life*, xiv, 7–9, transl. Bernadotte Perrin)

Not only did Archimedes overturn power relations through the intermediary of the compound pulley, he also reversed political relations by
offering the king a real mechanism for making one man physically
stronger than a multitude. Up to that time, the Sovereign represented the
masses whose spokesperson he was, but he had no greater strength as a
result. Archimedes procured a different principle of composition for the
Leviathan by transforming the relation of political representation into a
relation of mechanical proportion. Without geometry and statics, the
Sovereign had to reckon with social forces that infinitely overpowered
him. But if you add the lever of technology to the play of political
representation alone, then you can become stronger than the multitude;
you can attack and defend yourself. It is not surprising that Hiero was
'amazed' at the power of technology (*sunnoésas tès tecnès tén dunamin*).
It had not occurred to him, until then, to bring political power into
relation with the compound pulley.

But Plutarch's lesson goes still further. This first moment through
which Archimedes makes (physical) force commensurable with (political)
force owing to the relation of proportion between large and small,
between the reduced model and the life-size application, is coupled with a
second, even more decisive moment:

> And yet, Archimedes [after equipping Syracuse with war machines]
> possessed such a lofty spirit, so profound a soul, and such a wealth of
> scientific theory, that although his inventions had won for him a name and
> fame for superhuman sagacity, he would not consent to leave behind him
> any treatise on this subject, but regarding the work of an engineer and
> every art that ministers to the needs of life as ignoble and vulgar, he
> devoted his earnest efforts only to those studies the subtlety and charm of
> which are not affected by the claims of necessity. (Plutarch, xvii, 4–5)

Mathematical demonstrations remain incommensurable with lowly
manual trades, vulgar politics, mere applications. Archimedes is divine,
the power of mathematics is supernatural. All vestiges of composition,
connection, alliance, liaison between the two moments are now effaced.
Even treatises have to disappear without trace. The first moment
produced an unknown hybrid thanks to which the weaker became the
stronger through the alliance he established between political forms and
the laws of proportion. The second moment purifies politics and science,
the empire of men and the empyrean of mathematics, and renders them
incomparable (Serres, 1989). The Archimedean point is to be sought not
in the first moment, but in the conjunction of the two: how are we to
undertake politics with new means rendered suddenly commensurable,

while rejecting any link between absolutely incommensurable activities? The balance sheet is doubly positive: Hiero defends Syracuse with the machines whose dimensions we know how to calculate through proportions, and the collective also grows proportionally; but the origin of this variation in scale, of this commensurability, disappears for ever, leaving the empyrean of mathematics as a resource of fresh forces, always available, never visible. Yes, science is indeed politics pursued by other means, means that are powerful only because they remain radically other (Latour, 1990b).

By learning of Archimedes' coup (or rather, Plutarch's) we identify the entry point of a new type of nonhumans into the very fabric of the collective. It is not a matter of trying to find out how geometry 'reflects' Hiero's interests, or how Syracusan society 'is constrained' by the laws of geometry. A new collective is constituted by enlisting geometry and denying that it has done so. Society cannot explain geometry, since it is a new geometry-based society that begins to defend the walls of Syracuse against Marcellus. Politics-based society is an artifact obtained by the elimination of walls and levers, pulleys and swords, just as the social context of seventeenth-century England could be obtained only by the preliminary exclusion of the air pump and the nascent science of physics. It is only when we remove the nonhumans churned up by the collective that the residue, which we call society, becomes incomprehensible, because its size, its durability and its solidity no longer have a cause. One might as well sustain the Leviathan with naked citizens and the social contract alone, without air pumps, sword, blade, invoices, computers, files and palaces (Callon and Latour, 1981; Latour, 1988c; Strum and Latour, 1987). The social link does not hold without the objects that the other branch of the Constitution permits us both to mobilize and to render forever incommensurable with the social world.

4.8 Absolute Relativism and Relativist Relativism

The question of relativism is not closed, however, even if we take into account simultaneously the profound likeness of natures-cultures – the old anthropological matrix – and the difference in size, the scope of the mobilization of these collectives. In fact, as I have indicated several times, size is related to the modern Constitution. It is precisely because the Constitution guarantees that quasi-objects will be absolutely and irreversibly transformed, either into objects of external nature or into subjects of society, that the mobilization of these quasi-objects can take on an unprecedented amplitude. Symmetrical anthropology thus has to do justice to this peculiarity, without adding to it any epistemological break, any Great Metaphysical Divide, any difference between prelogical and logical societies, 'hot' ones and 'cold' ones, between an Archimedes who meddles in politics and a divine Archimedes with his head in the celestial Heavens of Ideas. The whole challenge of the exercise is to generate a maximum of differences by a minimum of means (Goody, 1977; Latour, 1990a).

Moderns do differ from premoderns by this single trait: they refuse to conceptualize quasi-objects as such. In their eyes, hybrids present the horror that must be avoided at all costs by a ceaseless, even maniacal purification. By itself, this difference in constitutional representation would not matter very much, since it would not suffice to set moderns apart from others. There are as many purification processes as there are collectives. But the machine for creating differences is triggered by the refusal to conceptualize quasi-objects, because this very refusal leads to the uncontrollable proliferation of a certain type of being: *the object, constructor of the social, expelled from the social world, attributed to a transcendent world that is, however, not divine – a world that produces, in contrast, a floating subject, bearer of law and morality.* Boyle's air pump, Pasteur's microbes, Archimedes' pulleys, are such objects. These new nonhumans possess miraculous properties because they are at one and the same time both social and asocial, producers of natures and constructors of subjects. They are the tricksters of comparative anthropology. Through this opening, sciences and technologies will emerge in society in such a mysterious way that this miracle will force Westerners to see themselves as completely different from others. The first miracle gives rise to a second (why don't the others do the same?), then a third (why are we so exceptional?). This feature generates a cascade of small differences that will be collected, summarized and amplified by the Great Divide, the great narrative of the West, set radically apart from all cultures.

Once this feature has been pinpointed, and thereby neutralized, relativism offers no more significant difficulties. Nothing keeps us from reopening the question of how to establish relationships among collectives by defining two relativisms that have hitherto been conflated. The first is absolute; the second is relative. The first locked cultures away in exoticism and strangeness, because it accepted the universalists' viewpoint while refusing to rally round it: if no common, unique and transcendental measuring instrument exists, then all languages are untranslatable, all intimate emotions incommunicable, all rites equally respectable, all paradigms incommensurable. There is no arguing about tastes or colours. Whereas universalists declare that this common yardstick does exist, absolute relativists are delighted that there is no such thing. Their attitudes may differ, but both groups agree in asserting that the reference to some absolute yardstick is essential to their dispute.

This amounts to not taking the practice of relativism, or even the word relativism, very seriously. To establish relations; to render them commensurable; to regulate measuring instruments; to institute metrological chains; to draw up dictionaries of correspondences; to discuss the compatibility of norms and standards; to extend calibrated networks; to set up and negotiate valorimeters – these are some of the meanings of the word 'relativism' (Latour, 1988d). Absolute relativism, like its enemy brother rationalism, forgets that measuring instruments have to be set up. By ignoring the work of instrumentation, by conflating science with nature, one can no longer understand anything about the notion of commensurability itself. They neglect even more thoroughly the enormous efforts Westerners have made to 'take the measure' of other peoples, to 'size them up' by rendering them commensurable and by creating measuring standards that did not exist before – via military and scientific expeditions.

But if we are to understand this task of measuring, we need to reinforce the noun with the adjective 'relativist', which compensates for the noun's apparent foolishness. Relativist relativism restores the compatibility that was assumed to have been lost. To be sure, relativist relativism has to abandon what constituted the common argument of the universalists as well as the earliest cultural relativists – that is, the word 'absolute'. Instead of stopping midway, it continues to the end and rediscovers, in the form of work and montage, practice and controversy, conquest and domination, the process of establishing relations. A little relativism distances us from the universal; a lot brings us back, but it is a universal in networks that has no more mysterious properties.

The universalists defined a single hierarchy. The absolute relativists made all hierarchies equal. The relativist relativists, more modest but more empirical, point out what instruments and what chains serve to create asymmetries and equalities, hierarchies and differences (Callon, 1992). Worlds appear commensurable or incommensurable only to those who cling to measured measures. Yet all measures, in hard and soft science alike, are also measuring measures, and they construct a commensurability that did not exist before their own calibration. Nothing is, by itself, either reducible or irreducible to anything else. Never by itself, but always through the mediation of another. How can one claim that worlds are untranslatable, when translation is the very soul of the process of relating? How can one say that worlds are dispersed, when there are hundreds of institutions that never stop totalizing them? Anthropology itself – one discipline among many others, one institution among many others – participates in the work of relating, of constructing catalogues and museums, of sending missions, expeditions and investigators, maps, questionnaires, and filing systems (Copans and Jamin, 1978; Fabian, 1983; Stocking, 1983, 1986). Ethnology is one of those measuring measures that resolves the question of relativism in practical terms by constructing a certain commensurability. If the question of relativism is insoluble, relativist relativism – or, to put it more elegantly, relationism – presents no difficulty in principle. If we cease to be completely modern, relationism will become one of the essential resources for relating the collectives that will no longer be targets for modernization. Relationism will serve as an organon for planetary negotiations over the relative universals that we are groping to construct.

4.9 Small Mistakes Concerning the Disenchantment of the World

We are indeed different from others, but we must not situate the differences where the now-closed question of relativism had located them. As collectives, we are all brothers. Except in the matter of dimension, which is itself caused by small differences in the distribution of entities, we can recognize a continuous gradient between premoderns and nonmoderns. Unfortunately, the difficulty of relativism does not arise only from the bracketing off of Nature. It stems also from the related belief that the modern world is truly disenchanted. It is not only out of arrogance that Westerners think they are radically different from others, it is also out of despair, and by way of self-punishment. They like to frighten themselves with their own destiny. Their voices quaver when they contrast Barbarians to Greeks, or the Centre to the Periphery, or when they celebrate the Death of God, or the Death of Man, the European *Krisis*, imperialism, anomie, or the end of the civilizations that we now know are mortal. Why do we get so much pleasure out of being so different not only from others but from our own past? What psychologist will be subtle enough to explain our morose delight in being in perpetual crisis and in putting an end to history? Why do we like to transform small differences in scale among collectives into huge dramas?

In order to bypass completely the modern pathos that prevents us from recognizing the fraternity of collectives, and thus to sort them more freely, comparative anthropology has to measure these effects of size with precision. Now the modern Constitution requires that the scaling effects of our collectives be confused with their causes, which the Constitution cannot indicate without ceasing to be operative. Rightly astounded by the size of the effects, the moderns believe that they require prodigious causes. And as the only causes recognized by the Constitution appear miraculous because they are reversed, the moderns clearly have to imagine themselves as different from ordinary humanity. In their hands, the uprooted, acculturated, Americanized, scientified, technologized Westerner becomes a Spock-like mutant. Haven't we shed enough tears over the disenchantment of the world? Haven't we frightened ourselves enough with the poor European who is thrust into a cold soulless cosmos, wandering on an inert planet in a world devoid of meaning? Haven't we shivered enough before the spectacle of the mechanized proletarian who is subject to the absolute domination of a mechanized capitalism and a Kafkaesque bureaucracy, abandoned smack in the middle of language games, lost in cement and formica? Haven't we felt sorry enough for the consumer who leaves the driver's seat of his car only to move to the sofa in the TV room where he is manipulated by the powers of the media and the postindustrialized society?! How we do love to wear the hair shirt of the absurd, and what even greater pleasure we take in postmodern nonsense!

However, we have never abandoned the old anthropological matrix. We have never stopped building our collectives with raw materials made of poor humans and humble nonhumans. How could we be capable of disenchanting the world, when every day our laboratories and our factories populate the world with hundreds of hybrids stranger than those of the day before? Is Boyle's air pump any less strange than the Arapesh spirit houses (Tuzin, 1980)? Does it contribute any less to constructing seventeenth-century England? How could we be victims of reductionism, when each scientist multiplies new entities by the thousands in order to be reductionist for a few of them? How could we be rationalists, when we still don't see beyond the tip of our own noses? How could we be materialists, when every matter we invent possesses new properties that no single matter allows us to unify (Dagognet, 1989)? How could we be victims of a total technological system, when machines are made of subjects and never succeed in settling into more or less stable systems (Kidder, 1981; Latour, 1992a)? How could we be chilled by the cold breath of the sciences, when the sciences are hot and fragile, human and controversial, full of thinking reeds and of subjects who are themselves inhabited by things (Pickering, 1992)?

The error the moderns make about themselves is easy enough to understand, once symmetry has been reestablished and once both the work of purification and the work of translation have been taken into account. The moderns confused products with processes. They believed that the production of bureaucratic rationalization presupposed rational bureaucrats; that the production of universal science depended on universalist scientists; that the production of effective technologies led to the effectiveness of engineers; that the production of abstraction was itself abstract; that the production of formalism was itself formal. We might just as well say that a refinery produces oil in a refined manner, or that a dairy produces butter in a butterly way! The words 'science', 'technology', 'organization', 'economy', 'abstraction', 'formalism', and 'universality' designate many real effects that we must indeed respect and for which we have to account. But in no case do they designate the causes of these same effects. These words are good nouns, but they make lousy adjectives and terrible adverbs. Science does not produce itself scientifically any more than technology produces itself technologically or economy economically. Scientists in the lab, Boyle's descendants, know this perfectly well, but as soon as they set out to reflect on what they do, they pronounce the words that sociologists and epistemologists, Hobbes's descendants, put in their mouths.

The paradox of the moderns (and the antimoderns) is that from the outset they have accepted massive cognitive or psychological explanations in order to explain equally massive effects, whereas in all other scientific domains they seek small causes for large effects. Reductionism has never been applied to the modern world, whereas it was supposed to have been applied to everything! Our own mythology consists in imagining ourselves as radically different, even before searching out small differences and small divides. However, as soon as the double Great Divide disappears, this mythology unravels as well. As soon as the work of mediation is taken into account simultaneously with the work of purification, ordinary humanity and ordinary inhumanity must come back in. To our great surprise, we then discover that we know very little about what causes sciences, technologies, organizations and economies. Open books on social science and epistemology, and you will see how they use the adjectives and adverbs 'abstract', 'rational', 'systematic', 'universal', 'scientific', 'organized', 'total', 'complex'. Look for the ones that try to explain the nouns 'abstraction', 'rationality', 'system', 'universal', 'science', 'organization', 'totality', 'complexity', without ever using the corresponding adjectives, or the equivalent adverbs, and you will be lucky to find a dozen. Paradoxically, we know more about the Achuar, the Arapesh or the Alladians than we know about ourselves. As long as small local causes lead to local differences, we are able to follow them. Why would we no longer be capable of following the thousand paths, with their strange topology, that lead from the local to the global and return to the local? Is anthropology forever condemned to be reduced to territories, unable to follow networks?

4.10 Even a Longer Network Remains Local at All Points

To take the precise measure of our differences without reducing them as relativism used to do, and without exaggerating them as modernizers tend to do, let us say that the moderns have simply invented longer networks by enlisting a certain type of nonhumans. The network-lengthening process had been interrupted in earlier periods, because it would have threatened the maintenance of territories (Deleuze and Guattari, [1972] 1983). But by multiplying the hybrids, half object and half subject, that we call machines and facts, collectives have changed their topography. Since this enlistment of new beings had enormous scaling effects by causing relations to vary from local to global, but we continue to think about them in terms of the old opposite categories of universal and contingent, we tend to transform the lengthened networks of Westerners into systematic and global totalities. To dispel this mystery, it suffices to follow the unaccustomed paths that allow this variation in scale, and to look at networks of facts and laws rather as one looks at gas lines or sewage pipes.

The secular explanation of the effects of size proper to the West is easy to grasp in technological networks (Bijker and others, 1987). If relativism had been applied there first, it would have had no trouble understanding this relative universal that is its greatest claim to glory. Is a railroad local or global? Neither. It is local at all points, since you always find sleepers and railroad workers, and you have stations and automatic ticket machines scattered along the way. Yet it is global, since it takes you from Madrid to Berlin or from Brest to Vladivostok. However, it is not universal enough to be able to take you just anywhere. It is impossible to reach the little Auvergnat village of Malpy by train, or the little Staffordshire village of Market Drayton. There are continuous paths that lead from the local to the global, from the circumstantial to the universal, from the contingent to the necessary, only so long as the branch lines are paid for.

The railroad model can be extended to all the technological networks that we encounter daily. It may be that the telephone has spread everywhere, but we still know that we can die right next to a phone line if we aren't plugged into an outlet and a receiver. The sewer system may be comprehensive, but nothing guarantees that the tissue I drop on my bedroom floor will end up there. Electromagnetic waves may be everywhere, but I still have to have an antenna, a subscription and a decoder if I am to get CNN (Cable News Network). Thus, in the case of technological networks, we have no difficulty reconciling their local aspect and their global dimension. They are composed of particular places, aligned by a series of branchings that cross other places and require other branchings in order to spread. Between the lines of the network there is, strictly speaking, nothing at all: no train, no telephone, no intake pipe, no television set. Technological networks, as the name indicates, are nets thrown over spaces, and they retain only a few scattered elements of those spaces. They are connected lines, not surfaces. They are by no means comprehensive, global or systematic, even though they embrace surfaces without covering them, and extend a very long way. The work of relative universalization remains an easy-to-grasp category that relationism can follow in a thoroughgoing way. Every branching, every alignment, every connection can be documented, since it generates tracers, and every one of them has a cost. It can be extended almost everywhere; it can be spread out in time as well as in space, yet without filling time and space (Stengers, 1983).

For ideas, knowledge, laws, and skills, however, the model of the technological network seems inadequate to those who are highly impressed by the effects of diffusion, those who believe what epistemology says about the sciences. The tracers become more difficult to follow, their cost is no longer so well documented, and one risks losing sight of the bumpy path that leads from the local to the global. So the ancient philosophical category of the universal radically different from the contingent circumstances is applied to them.

It seems, then, that ideas and knowledge can spread everywhere without cost. Certain ideas appear to be local, others global. Universal gravitation appears to be active and present everywhere; we are convinced of it. Boyle's laws, Mariotte's laws, Planck's constants legislate everywhere and are constant everywhere. As for Pythagoras' theorem and transfinite numbers, they seem so universal that they may even escape this world here below to rejoin the works of the divine Archimedes. It is here that the old relativism and its enemy brother rationalism begin to show their faces, since it is in relation to these universals, and only these, that the humble Achuar or the poor Arapesh or the unfortunate Burgundians appear desperately contingent and arbitrary, forever imprisoned within the narrow confines of their regional peculiarities and their local knowledge (Geertz, 1971). If we had had only the world-economies of the Venetian, Genoan or American merchants, if we had had only telephones and television, railroads and sewers, Western domination would never have appeared as anything but the provisional and fragile extension of some frail and tenuous networks. But there is science, which always renews and totalizes and fills the gaping holes left by the networks in order to turn them into sleek, unified surfaces that are absolutely universal. Only the idea that we have had of science up to now rendered absolute a dominion that might have remained relative. All the subtle pathways leading continuously from circumstances to universals have been broken off by the epistemologists, and we have found ourselves with pitiful contingencies on one side and necessary Laws on the other – without, of course, being able to conceptualize their relations.

Now, as concepts, 'local' and 'global' work well for surfaces and geometry, but very badly for networks and topology. The belief in rationalization is a simple category mistake. One branch of mathematics has been confused with another! The itinerary of ideas, knowledge or facts would have been understood with no trouble if we had treated them like technological networks (Schaffer, 1988, 1991; Shapin and Schaffer, 1985; Warwick, 1992). Fortunately, the assimilation is made easier not only by the end of epistemology but also by the end of the Constitution, and by the technological transformations that it authorizes without including them. The itinerary of facts becomes as easy to follow as that of railways or telephones, thanks to the materialization of the spirit that thinking machines and computers allow. When information is measured in bytes and bauds, when one subscribes to a data bank, when one can plug into (or unplug from) a network of distributed intelligence, it is harder to go on picturing universal thought as a spirit hovering over the waters (Lévy, 1990). Reason today has more in common with a cable television network than with Platonic ideas. It thus becomes much less difficult than it was in the past to see our laws and our constants, our demonstrations and our theorems, as stabilized objects that circulate widely, to be sure, but remain within well-laid-out metrological networks from which they are incapable of exiting – except through branchings, subscriptions and decodings.

To speak in popular terms about a subject that has been dealt with largely in learned discourse, we might compare scientific facts to frozen fish: the cold chain that keeps them fresh must not be interrupted, however briefly. The universal in networks produces the same effects as the absolute universal, but it no longer has the same fantastic causes. It is possible to verify gravitation 'everywhere', but at the price of the relative extension of the networks for measuring and interpreting. The air's spring can be verified everywhere, provided that one hooks up to an air pump that spreads little by little throughout Europe owing to the multiple transformations of the experimenters (Shapin and Schaffer, 1985). Try to verify the tiniest fact, the most trivial law, the humblest constant, without subscribing to the multiple metrological networks, to laboratories and instruments. The Pythagorean theorem and Planck's constant spread into schools and rockets, machines and instruments, but they do not exit from their worlds any more than the Achuar leave their villages. The former constitute lengthened networks, the latter territories or loops: the difference is important and must be respected, but let us not use it to justify transforming the former into universals and the latter into localities. To be sure, the West may believe that universal gravitation is universal even in the absence of any instrument, any calculation, any decoding, any laboratory, just as the Bimin-Kuskumin of New Guinea may believe that they comprise all of humanity, but these are respectable beliefs that comparative anthropology is no longer obliged to share.

4.11 The Leviathan is a Skein of Networks

Just as the moderns have been unable to keep from exaggerating the universality of their sciences (by pulling away the subtle network of practices, instruments and institutions that paved the way from contingencies to necessities), symmetrically, they have been unable to do anything but exaggerate the size and solidity of their own societies. They thought themselves revolutionary because they invented the universality of sciences that were torn out of local peculiarities for all time, and because they invented gigantic rationalized organizations that broke with all the local loyalties of the past. In so doing, they missed the originality of their own inventions twice over: a new topology that makes it possible to go almost everywhere, yet without occupying anything except narrow lines of force and a continuous hybridization between socialized objects and societies rendered more durable through the proliferation of nonhumans. The moderns got excited about virtues they are incapable of possessing (rationalization), but they likewise flagellated themselves for sins they are quite incapable of committing (rationalization again)! In both cases, they mistook length or connection for differences in level. They thought there really were such things as people, ideas, situations that were local and organizations, laws, rules that were global. They believed that there were contexts and other situations that enjoyed the mysterious property of being 'decontextualized' or 'delocalized'. And indeed, if the intermediary network of quasi-objects is not reconstituted, it becomes just as difficult to grasp society as scientific truth, and for the same reasons. The mediators that have been effaced had contained everything, while the extremes, once isolated, are no longer anything at all.

Without the countless objects that ensured their durability as well as their solidity, the traditional objects of social theory – empire, classes, professions, organizations, States – become so many mysteries (Law, 1986, 1992; Law and Fyfe, 1988). What, for example, is the size of IBM, or the Red Army, or the French Ministry of Education, or the world market? To be sure, these are all actors of great size, since they mobilize hundreds of thousands or even millions of agents. Their amplitude must therefore stem from causes that absolutely surpass the small collectives of the past. However, if we wander about inside IBM, if we follow the chains of command of the Red Army, if we inquire in the corridors of the Ministry of Education, if we study the process of selling and buying a bar of soap, we never leave the local level. We are always in interaction with four or five people; the building superintendent always has his territory well staked out; the directors' conversations sound just like those of the employees; as for the salespeople, they go on and on giving change and filling out their invoices. Could the macro-actors be made up of micro-actors (Garfinkel, 1967)? Could IBM be made up of a series of local interactions? The Red Army of an aggregate of conversations in the mess hall? The Ministry of Education of a mountain of pieces of paper? The world market of a host of local exchanges and arrangements?

We rediscover the same problem as that of trains, telephones, or universal constants. How can one be connected without being either local or global? Modern sociologists and economists have a hard time posing the problem. Either they remain at the 'micro' level, that of interpersonal contacts, or they move abruptly to the 'macro' level and no longer deal with anything, they believe, but decontextualized and depersonalized rationalities. The myth of the soulless, agentless bureaucracy, like that of the pure and perfect marketplace, offers the mirror-image of the myth of universal scientific laws. Instead of the continual progression of an inquiry, the moderns have imposed an ontological difference as radical as the sixteenth-century differentiation between the supralunar worlds that knew neither change nor uncertainty. (The same physicists had a good laugh with Galileo at that ontological distinction – but then they rushed to reestablish it in order to protect the laws of physics from social corruption!)

Yet there is an Ariadne's thread that would allow us to pass with continuity from the local to the global, from the human to the nonhuman. It is the thread of networks of practices and instruments, of documents and translations. An organization, a market, an institution, are not supralunar objects made of a different matter from our poor local sublunar relations (Cambrosio *et al.* 1990). The only difference stems from the fact that they are made up of hybrids and have to mobilize a great number of objects for their description. The capitalism of Karl Marx or Fernand Braudel is not the total capitalism of the Marxists (Braudel, 1985). It is a skein of somewhat longer networks that rather inadequately embrace a world on the basis of points that become centres of profit and calculation. In following it step by step, one never crosses the mysterious *limes* that should divide the local from the global. The organization of American big business described by Alfred Chandler (Chandler, 1977, 1990) is not the Organization described by Kafka. It is a braid of networks materialized in order slips and flow charts, local procedures and special arrangements, which permit it to spread to an entire continent so long as it does not cover that continent. One can follow the growth of an organization in its entirety without ever changing levels and without ever discovering 'decontextualized' rationality. The very size of a totalitarian State is obtained only by the construction of a network of statistics and calculations, of offices and inquiries, which in no way corresponds to the fantastic topography of the total State (Desrosières, 1990). The scientifico-technological empire of Lord Kelvin described by Norton Wise (Smith and Wise, 1989), or the electricity market as described by Tom Hughes (Hughes, 1983), never require us to leave the particularities of the laboratory, the meeting room or the control centre. Yet these 'networks of power' and these 'lines of force' do extend across the entire world. The markets described by the Economy of conventions are indeed regulated and global, even though none of the causes of that regulation and that aggregation is itself either global or total. The aggregates are not made from some substance different from what they are aggregating (Thévenot, 1989, 1990). No visible or invisible hand suddenly descends to bring order to dispersed and chaotic individual atoms. The two extremes, local and global, are much less interesting than the intermediary arrangements that we are calling networks.

4.12 A Perverse Taste for the Margins

Just as the adjectives 'natural' and 'social' designate representations of collectives that are neither natural nor social in themselves, so the words 'local' and 'global' offer points of view on networks that are by nature neither local nor global, but are more or less long and more or less connected. What I have called modern exoticism consists in taking these two pairs of oppositions as what defines our world and what would set us apart from all others. So four different regions are thus created. The natural and the social are not composed of the same ingredients; the global and the local are intrinsically distinct. Yet we know nothing about the social that is not defined by what we think we know about the natural, and vice versa. Similarly, we define the local only by contrast with what we think we have to attribute to the global, and vice versa. So the strength of the error that the modern world makes about itself is now understandable, when the two couples of opposition are paired: in the middle there is nothing thinkable – no collective, no network, no mediation; all conceptual resources are accumulated at the four extremes. We poor subject-objects, we humble societies-natures, we modest locals-globals, are literally quartered among ontological regions that define each other mutually but no longer resemble our practices.

This quartering makes it possible to unfurl the tragedy of modern man considering himself as absolutely and irremediably different from all other humanities and all other naturalities. But such a tragedy is not inevitable, if we recall that these four terms are representations without any direct relation to the collectives and the networks that give them meaning. In the middle, where nothing is supposed to be happening, there is almost everything. And at the extremes – which according to the moderns house the origin of all forces, Nature and Society, Universality and Locality – there is nothing except purified agencies that serve as constitutional guarantees for the whole.

The tragedy becomes more painful still when the antimoderns, taking what the moderns say about themselves at face value, want to save something from what looks to them like a shipwreck. The antimoderns firmly believe that the West has rationalized and disenchanted the world, that it has truly peopled the social with cold and rational monsters which saturate all of space, that it has definitively transformed the premodern cosmos into a mechanical interaction of pure matters. But instead of seeing these processes as the modernizers do – as glorious, albeit painful, conquests – the antimoderns see the situation as an unparalleled catastrophe. Except for the plus or minus sign, moderns and antimoderns share all the same convictions. The postmoderns, always perverse, accept the idea that the situation is indeed catastrophic, but they maintain that it is to be acclaimed rather than bemoaned! They claim weakness as their ultimate virtue, as one of them affirms in his own inimitable style: 'The *Verwindung* of metaphysics is exercised as *Verwindung* of the *Ge-Stell*' (Vatimo, 1987, p. 184).

What do the antimoderns do, then, when they are confronted with this shipwreck? They take on the courageous task of saving what can be saved: souls, minds, emotions, interpersonal relations, the symbolic dimension, human warmth, local specificities, hermeneutics, the margins and the peripheries. An admirable mission, but one that would be more admirable still if all those sacred vessels were actually threatened. Now where does the threat come from? Surely not from collectives incapable of abandoning their fragile and narrow networks populated with souls and objects. Surely not from sciences whose relative universality has to be purchased, day after day, by branchings and calibrations, instruments and alignments. Surely not from societies whose size varies only so long as material entities characterized by variable ontology proliferate. Where does it come from, then? Well, in part from the antimoderns themselves, and from their accomplices the moderns, who frighten each other and add gigantic causes to the effects of size. 'You are disenchanting the world; I shall maintain the rights of the spirit!' 'You want to maintain the spirit? Then we shall materialize it!' 'Reductionists!' 'Spiritualists!' The more the antireductionists, the romantics, the spiritualists seek to save subjects, the more the reductionists, the scientistics, the materialists imagine that they possess objects. The more the latter boast, the more they frighten the former; the wilder the former become, the more the latter believe that they themselves are indeed terrifying. Are not most ethicists busy with those two opposite but symmetrical tasks: defending the purity of science and rationality from the polluting influence of passions and interests; defending the unique values and rights of human subjects against the domination of scientific and technical objectivity?

The defence of marginality presupposes the existence of a totalitarian centre. But if the centre and its totality are illusions, acclaim for the margins is somewhat ridiculous. It is fine to want to defend the claims of the suffering body and human warmth against the cold universality of scientific laws. But if universality stems from a series of places in which warm flesh-and-blood bodies are suffering everywhere, is not this defence grotesque? Protecting human beings from the domination of machines and technocrats is a laudable enterprise, but if the machines are full of human beings who find their salvation there, such a protection is merely absurd (Ellul, 1967). It is admirable to demonstrate that the strength of the spirit transcends the laws of mechanical nature, but this programme is idiotic if matter is not at all material and machines are not at all mechanical. It is admirable to seek to save Being, with a cry of desperation, at the very moment when technological *Ge-Stell* seems to dominate everything, because 'where danger is, grows the saving power also'. But it is rather perverse to seek to profit brazenly from a crisis that has not yet commenced!

Look for the origins of the modern myths, and you will almost always find them among those who claim to be countering modernism with the impenetrable barrier of the spirit, of emotion, the subject, or the margins. In the effort to offer a supplement of soul to the modern world, the one it has is taken away – the one it had, the one it was quite incapable of losing. That subtraction and that addition are the two operations that allow the moderns and the antimoderns to frighten each other by agreeing on the essential point: we are absolutely different from the others, and we have broken radically with our own past. Now sciences and technologies, organizations and bureaucracies are the only proofs always offered by moderns and antimoderns of that unparalleled catastrophe, and it is precisely through them that science studies can demonstrate the permanence of the old anthropological matrix best and most directly. To be sure, the innovation of lengthened networks is important, but it is hardly a reason to make such a great fuss.

4.13 Avoid Adding New Crimes to Old

It is quite difficult, however, to soothe the modern sense of dereliction, because its starting point is a sentiment that is respectable in itself: the awareness of having committed irreparable crimes against the rest of the natural and cultural worlds, as well as crimes against the self whose scope and intentions seem indeed without precedent. How can moderns be restored to ordinary humanity and inhumanity without being too hastily absolved of the crimes that they are right to seek to expiate? How can we claim – correctly – that our crimes are frightful, but that they remain ordinary; that our virtues are great, but that they too are quite ordinary?

Our misdeeds can be compared to our access to Nature: we must not exaggerate their causes even as we measure their effects, for that exaggeration itself would be the cause of greater crimes. Every totalization, even if it is critical, helps totalitarianism. We need not add total domination to real domination. Let us not add power to force. We need not grant total imperialism to real imperialism. We need not add absolute deterritorialization to capitalism, which is also quite real enough (Deleuze and Guattari, [1972] 1983). Similarly, we do not need to credit scientific truth and technological efficacity with transcendence, also total, and rationality, also absolute. With misdeeds as with domination, with capitalisms as with sciences, what we need to understand is the ordinary dimension: the small causes and their large effects (Arendt, 1963; Mayer, 1988).

Demonizing may be more satisfying for us because we still remain exceptional even in evil; we remain cut off from all others and from our own past, modern at least for the worst after thinking we were modern for the best. But totalization participates, in devious ways, in what it claims to abolish. It renders its practitioners powerless in the face of the enemy, whom it endows with fantastic properties. A system that is total and sleek does not get divided up. A transcendental and homogeneous nature does not get recombined. A totally systematic technological system cannot be reshuffled by anyone. A Kafkaesque society cannot be renegotiated. A 'deterritorializing' and absolutely schizophrenic capitalism will never be redistributed by anyone. A West radically cut off from other cultures-natures is not open to discussion. Cultures imprisoned for ever in arbitrary, complete and consistent representations cannot be evaluated. A world that has totally forgotten Being will be saved by no one. A past from which we are forever separated by radical epistemological breaks cannot be sorted out again by anyone at all.

All these supplements of totality are attributed by their critics to actors who did not ask for them. Take some small business-owner hesitatingly going after a few market shares, some conqueror trembling with fever, some poor scientist tinkering in his lab, a lowly engineer piecing together a few more or less favourable relationships of force, some stuttering and fearful politician; turn the critics loose on them, and what do you get? Capitalism, imperialism, science, technology, domination – all equally absolute, systematic, totalitarian. In the first scenario, the actors were trembling; in the second, they are not. The actors in the first scenario could be defeated; in the second, they no longer can. In the first scenario, the actors were still quite close to the modest work of fragile and modifiable mediations; now they are purified, and they are all equally formidable.

What is to be done, then, with such sleek, filled-in surfaces, with such absolute totalities? Turn them inside out all at once, of course; subvert them, revolutionize them – such was the strategy of those modernists *par excellence*, the Marxists. Oh, what a lovely paradox! By means of the critical spirit, the moderns have invented at one and the same time the total system, the total revolution to put an end to the system, and the equally total failure to carry out that revolution – a failure that leaves them in total postmodern despair! Isn't this the cause of many of the crimes with which we reproach ourselves? By considering the Constitution instead of the work of translation, the critics have imagined that we were incapable of tinkering, reshuffling, crossbreeding and sorting. On the basis of the fragile heterogeneous networks that collectives have always formed, the critics have elaborated homogeneous totalities that could not be touched unless they were totally revolutionized. And because this subversion was impossible, but they tried it anyway, they have gone from one crime to another. How could the totalizers' '*Noli me tangere*' still be passed off as a proof of morality? Might the belief in a radical and total modernity then lead to immorality?

Perhaps it would be less unjust to speak of a generational effect. We were born after the war, with the black camps and then the red camps behind us, with famines below us, the nuclear apocalypse over our heads, and the global destruction of the planet ahead of us. It is indeed difficult for us to deny the effects of scale, but it is still more difficult to believe unhesitatingly in the incomparable virtues of the political, medical, scientific or economic revolutions. Yet we were born amid sciences, we have known only peace and prosperity, and we love – should we admit it? – the technologies and consumer objects that the philosophers and moralists of earlier generations advise us to abhor. For us, technologies are not new, they are not modern in the banal sense of the word, since they have always constituted our world. More than earlier generations, ours has digested, integrated, and perhaps socialized them. Because we are the first who believe neither in the virtues nor in the dangers of science and technology, but share their vices and virtues without seeing either heaven or hell in them, it is perhaps easier for us to look for their causes without appealing to the white man's burden, or the fatality of capitalism, or the destiny of Europe, or the history of Being, or universal rationality. Perhaps it is easier today to give up the belief in our own strangeness. We are not exotic but ordinary. As a result, the others are not exotic either. They are like us, they have never stopped being our brethren. Let us not add to the crime that of believing that we are radically different to all the others.

4.14 Transcendences Abound

If we are no longer entirely modern, and if we are not premodern either, then on what basis are we going to establish the comparison of collectives? As we now know, we have to add the unofficial work of mediation to the official Constitution. When we compared the Constitution to the cultures described by the asymmetrical anthropology of the past, we ended up only with relativism and an impossible modernization. If on the contrary, we compare the translation work of collectives, we make symmetrical anthropology possible, and we dispel the false problems of absolute relativism. But we also deprive ourselves of the resources developed by the moderns: the Social, Nature, Discourse – not to mention the crossed-out God. This is the ultimate difficulty of relativism: now that comparison has become possible, in what common space do all collectives, producers of natures and societies, find themselves equally immersed?

Are they in nature? Certainly not, since sleek, transcendent, external nature is the relative and belated consequence of collective production. Are they in society? Not there either, since society is only the symmetrical artifact of nature, what is left when all objects are removed, and the mysterious transcendence of the Leviathan is produced. Are they in language, then? Impossible, since discourse is another artifact that has meaning only when the external reality of the referent and the social context are both bracketed off. Are they in God? That is not very probable, for the metaphysical entity that bears this name merely occupies the place of a remote referee so as to maintain as much distance as possible between two symmetrical entities, Nature and Society. Are they in Being? That is even less likely since, through an astonishing paradox, the thought of Being has become precisely a residue, what is left over after every science, every technology, every society, every history,

every language, every theology, has been abandoned to the pure
expansionism of beings. Naturalization, socialization, discursivization,
divinization, ontologization – all these '-izations' are equally implausible.
None of them forms a common basis on which collectives, thus rendered
comparable, might repose. No, we do not fall from Nature into the
Social, from the Social into Discourse, from Discourse into God, from
God into Being. Those agencies had a constitutional role to play only so
long as they remained distinct. No one of them can cover, fill, subsume
the others; no one of them can serve to describe the work of mediation
and translation.

Where are we, then? Where do we land? As long as we keep asking
that question, we are unmistakably in the modern world, obsessed with
the construction of one immanence [*immanere*: to reside in] or the
deconstruction of another. We still remain – to use the old word – within
metaphysics. Now by traversing these networks, we do not come to rest
in anything particularly homogeneous. We remain, rather, within an
infra-physics. Are we immanent, then, one force among others, texts
among other texts, one society among other societies, being among
beings?

Not that either, for if, instead of attaching poor phenomena to the
solid hooks of Nature and Society, we let mediators produce natures and
societies, we reverse the direction of the modernizing transcendences.
Natures and societies become the relative products of history. However,
we do not fall into immanence alone, since networks are immersed in
nothing. We do not need a mysterious ether for them to propagate
themselves. We do not need to fill in blanks. It is the conception of the
terms 'transcendence' and 'immanence' that ends up being modified by
the moderns' return to nonmodernity. Who told us that transcendence
had to have a contrary? We have never abandoned transcendence – that
is, the maintenance in presence by the mediation of a pass.

Moderns were always struck by the diffuse aspect of active or spiritual
forces in other so-called premodern cultures. Nowhere were pure
matters, pure mechanical forces, put into play. Spirits and agents, gods
and ancestors, were blended in at every point. In contrast, from the
moderns' viewpoint the modern world appeared disenchanted, drained
of its mysteries, dominated by the sleek forces of pure immanence on
which we humans alone imposed some symbolic dimension and beyond
which there existed, perhaps, the transcendence of the crossed-out God.
Now if there is no immanence, if there are only networks, agents, actants,
we cannot be disenchanted. Humans are not the ones who arbitrarily add
the 'symbolic dimension' to pure material forces. These forces are as
transcendent, active, agitated, spiritual, as we are. Nature is no more
immediately accessible than society or the crossed-out God. Instead of

the subtle play of the moderns among three entities each of which was at once transcendent and immanent, we get a single proliferation of transcendences. A polemical term invented to counter the supposed invasion of immanence, the word has to change meaning if there is no longer an opposite term.

I call this transcendence that lacks a contrary 'delegation'. The utterance, or the delegation, or the sending of a message or a messenger, makes it possible to remain in presence – that is, to exist. When we abandon the modern world, we do not fall upon someone or something, we do not land on an essence, but on a process, on a movement, a passage – literally a pass, in the sense of this term as used in ball games. We start from a continuous and hazardous existence – continuous because it is hazardous – and not from an essence; we start from a presenting, and not from permanence. We start from the *vinculum* itself, from passages and relations, not accepting as a starting point any being that does not emerge from this relation that is at once collective, real and discursive. We do not start from human beings, those latecomers, nor from language, a more recent arrival still. The world of meaning and the world of being are one and the same world, that of translation, substitution, delegation, passing. We shall say that any other definition of essence is 'devoid of meaning'; in fact, it is devoid of the means to remain in presence, to last. All durability, all solidity, all permanence will have to be paid for by its mediators. It is this exploration of a transcendence without a contrary that makes our world so very ummodern, with all those nuncios, mediators, delegates, fetishes, machines, figurines, instruments, representatives, angels, lieutenants, spokespersons and cherubim. What sort of world is it that obliges us to take into account, at the same time and in the same breath, the nature of things, technologies, sciences, fictional beings, religions large and small, politics, jurisdictions, economies and unconsciousnesses? Our own, of course. That world ceased to be modern when we replaced all essences with the mediators, delegates and translators that gave them meaning. That is why we do not yet recognize it. It has taken on an ancient aspect, with all those delegates, angels and lieutenants. Yet it does not resemble the cultures studied by ethnologists, either, for Western ethnologists had never undertaken the symmetrical work of bringing delegates, mediators and translators back home, into their own community. Anthropology had been built on the basis of science, or on the basis of society, or on the basis of language; it always alternated between universalism and cultural relativism, and in the end it may have taught us as little about 'Them' as about 'Us'.

5

□

REDISTRIBUTION

5.1 The Impossible Modernization

After sketching out the modern Constitution and the reasons it had been invincible for so long; after showing why the critical revolution had been overwhelmed by the emergence of quasi-objects that obliged us to see the modern together with the nonmodern dimension; after reestablishing symmetry among collectives and thus measuring their differences in size while settling the question of relativism at the same time, I can now conclude this essay by tackling the most difficult question: the question of the nonmodern world that we are entering, I maintain, without ever having really left it.

Modernization, although it destroyed the near-totality of cultures and natures by force and bloodshed, had a clear objective. Modernizing finally made it possible to distinguish between the laws of external nature and the conventions of society. The conquerors undertook this partition everywhere, consigning hybrids either to the domain of objects or to that of society. The process of partitioning was accompanied by a coherent and continuous front of radical revolutions in science, technology, administration, economy and religion, a veritable bulldozer operation behind which the past disappeared for ever, but in front of which, at least, the future opened up. The past was a barbarian medley; the future, a civilizing distinction. To be sure, the moderns have always recognized that they too had blended objects and societies, cosmologies and sociologies. But this was in the past, while they were still only premodern. By increasingly terrifying revolutions, they have been able to tear themselves away from that past. Since other cultures still mix the constraints of rationality with the needs of their societies, they have to be helped to emerge from that confusion by annihilating their past.

Modernizers know perfectly well that even in their own midst islands of barbarianism remain, in which technological efficacity and social arbitrariness are excessively intertwined. But before long they will have achieved modernization, they will have liquidated those islands, and we shall all inhabit the same planet; we shall all be equally modern, all equally capable of profiting from what, alone, forever escapes the tyranny of social interest: economic rationality, scientific truth, technological efficiency.

Certain modernizers continue to speak as if such a fate were possible and desirable. However, one has only to express it to see how self-contradictory this claim is. How could we bring about the purification of sciences and societies at last, when the modernizers themselves are responsible for the proliferation of hybrids thanks to the very Constitution that makes them proliferate by denying their existence? For a long time, this contradiction was hidden by the moderns' very increase. Permanent revolutions in the State, and sciences, and technologies, were supposed to end up absorbing, purifying and civilizing the hybrids by incorporating them either into society or into nature. But the double failure that was my starting point, that of socialism – at stage left – and that of naturalism – at stage right – has made the work of purification less plausible and the contradiction more visible. There are no more revolutions in store to impel a continued forward flight. There are so many hybrids that no one knows any longer how to lodge them in the old promised land of modernity. Hence the postmoderns' abrupt paralysis.

Modernization was ruthless toward the premoderns, but what can we say about postmodernization? Imperialist violence at least offered a future, but sudden weakness on the part of the conquerors is far worse for, always cut off from the past, it now also breaks with the future. Having been slapped in the face with modern reality, poor populations now have to submit to postmodern hyperreality. Nothing has value; everything is a reflection, a simulacrum, a floating sign; and that very weakness, they say, may save us from the invasion of technologies, sciences, reasons. Was it really worth destroying everything to end up adding this insult to that injury? The empty world in which the postmoderns evolve is one they themselves, and they alone, have emptied, because they have taken the moderns at their word. Postmodernism is a symptom of the contradiction of modernism, but it is unable to diagnose this contradiction because it shares the same upper half of the Constitution – the sciences and the technologies are extrahuman – but it no longer shares the cause of the Constitution's strength and greatness – the proliferation of quasi-objects and the multiplication of intermediaries between humans and nonhumans allowed by the absolute distinction between humans and nonhumans.

However, the diagnosis is not very difficult to make, now that we are obliged to consider the work of purification and the work of mediation symmetrically. Even at the worst moments of the Western imperium, it was never a matter of clearly separating the Laws of Nature from social conventions once and for all. It was always a matter of constructing collectives by mixing a certain type of nonhumans and a certain type of humans, and extracting in the process Boyle-style objects and Hobbes-style subjects (not to mention the crossed-out God) on an ever-increasing scale. The innovation of longer networks is an interesting peculiarity, but it is not sufficient to set us radically apart from others, or to cut us off for ever from our past. Modernizers are not obliged to continue their revolutionary task by gathering their forces, ignoring the postmoderns' predicament, gritting their teeth, and continuing to believe in the dual promises of naturalism and socialism no matter what, since that particular modernization has never got off the ground. It was never anything but the official representation of another much more profound and different work that had always been going on and continues today on an ever-increasing scale. Nor are we obliged to struggle against modernization – in the militant manner of the antimoderns or the disillusioned manner of the postmoderns – since we would then be attacking the upper half of the Constitution alone, which we would merely be reinforcing while remaining unaware of what has always been the source of its vitality.

But does this diagnosis allow any remedy for the impossible moderniz-ation? If, as I have been saying all along, the Constitution allows hybrids to proliferate because it refuses to conceptualize them as such, then it remains effective only so long as it denies their existence. Now, if the fruitful contradiction between the two parts – the official work of purification and the unofficial work of mediation – becomes clearly visible, won't the Constitution cease to be effective? Won't moderniz-ation become impossible? Are we going to become – or go back to being – premodern? Do we have to resign ourselves to becoming antimodern? For lack of any better option, are we going to have to continue to be modern, but without conviction, in the twilight zone of the postmods?

5.2 Final Examinations

To answer these questions, we must first sort out the various positions I have outlined in the course of this essay, to bring the nonmodern to terms with the best those positions have to offer. What are we going to retain from the moderns? Everything, apart from exclusive confidence in the upper half of their Constitution, because this Constitution will need to be amended somewhat to include its lower half too. The moderns' greatness stems from their proliferation of hybrids, their lengthening of a certain type of network, their acceleration of the production of traces, their multiplication of delegates, their groping production of relative universals. Their daring, their research, their innovativeness, their tinkering, their youthful excesses, the ever-increasing scale of their action, the creation of stabilized objects independent of society, the freedom of a society liberated from objects – all these are features we want to keep. On the other hand, we cannot retain the illusion (whether they deem it positive or negative) that moderns have about themselves and want to generalize to everyone: atheist, materialist, spiritualist, theist, rational, effective, objective, universal, critical, radically different from other communities, cut off from a past that is maintained in a state of artificial survival due only to historicism, separated from a nature on which subjects or society would arbitrarily impose categories, denouncers always at war with themselves, prisoners of an absolute dichotomy between things and signs, facts and values.

Westerners felt far removed from the premoderns because of the External Great Divide – a simple exportation, as I have noted, of the Internal Great Divide. When the latter is dissolved, the former disappears, to be replaced by differences in size. Symmetrical anthropology has redistributed the Great Divide. Now that we are no longer so far removed from the premoderns – since when we talk about the premoderns we have to include a large part of ourselves – we are going to have to sort them out as well. Let us keep what is best about them, above all: the premoderns' inability to differentiate durably between the networks and the pure poles of Nature and Society, their obsessive interest in thinking about the production of hybrids of Nature and Society, of things and signs, their certainty that transcendences abound, their capacity for conceiving of past and future in many ways other than progress and decadence, the multiplication of types of nonhumans different from those of the moderns. On the other hand, we shall not retain the set of limits they impose on the scaling of collectives, localization by territory, the scapegoating process, ethnocentrism, and finally the lasting nondifferentiation of natures and societies.

But the sorting seems impossible and even contradictory in the face of what I have said above. Since the invention of longer networks and the increase in size of some collectives depends on the silence they maintain about quasi-objects, how can I promise to keep the changes of scale and give up the invisibility that allows them to spread? Worse still, how could I reject from the premoderns the lasting nondifferentiation of natures and societies, and reject from the moderns the absolute dichotomy between natures and societies? How can size, exploration, proliferation be maintained while the hybrids are made explicit? Yet this is precisely the amalgam I am looking for: *to retain the production of a nature and of a society that allow changes in size through the creation of an external truth and a subject of law, but without neglecting the co-production of sciences and societies.* The amalgam consists in using the premodern categories to conceptualize the hybrids, while retaining the moderns' final outcome of the work of purification – that is, an external Nature distinct from subjects. I want to keep following the gradient that leads from unstable existences to stabilized essences – and vice versa. To accomplish the work of purification, but as a particular case of the work of mediation. To maintain all the advantages of the moderns' dualism without its disadvantages – the clandestineness of the quasi-objects. To keep all the advantages of the premoderns' monism without tolerating its limits – the restriction of size through the lasting confusion of knowledge and power.

The postmoderns have sensed the crisis of the moderns and attempted to overcome it; thus they too warrant examination and sorting. It is of course impossible to conserve their irony, their despair, their discouragement, their nihilism, their self-criticism, since all those fine qualities depend on a conception of modernism that modernism itself has never really practised. As soon, however, as we add the lower part of the Constitution to the upper part, many of the intuitions of postmodernism are vindicated. For instance, we can save deconstruction – but since it no longer has a contrary, it turns into constructivism and no longer goes hand in hand with self-destruction. We can retain the deconstructionists' refusal of naturalization – but since Nature itself is no longer natural, this refusal no longer distances us from the sciences but, on the contrary, brings us closer to sciences in action. We can keep the postmoderns' pronounced taste for reflexivity – but since that property is shared among all the actors, it loses its parodic character and becomes positive. Finally, we can go along with the postmoderns in rejecting the idea of a coherent and homogeneous time that would advance by goose steps – but without retaining their taste for quotation and anachronism which maintains the belief in a truly surpassed past. Take away from the postmoderns their illusions about the moderns, and their vices become virtues – nonmodern virtues!

Regrettably, in the antimoderns I see nothing worth saving. Always on the defensive, they consistently believed what the moderns said about themselves and proceeded to affix the opposite sign to each declaration. Antirevolutionary, they held the same peculiar views as the moderns about time past and tradition. The values they defended were never anything but the residue left by their enemies; they never understood that the moderns' greatness stemmed, in practice, from the very reverse of

	What is retained	What is rejected
From the moderns	–long networks –size –experimentation –relative universals –final separation between objective nature and free society	–separation between nature and society –clandestineness of the practices of mediation –external Great Divide –critical denunciation –universality, rationality
From the premoderns	–non-separability of things and signs –transcendence without a contrary –multiplication of nonhumans –temporality by intensity	–obligation always to link the social and natural orders –scapegoating mechanism • ethnocentrism • territory –limits on scale
From the post moderns	–multiple times –constructivism –reflexivity –denaturalization	–belief in modernism –critical deconstruction –ironic reflexivity –anachronism

Figure 5.1 What is retained and what is rejected

what the antimoderns attacked them for. Even in their rearguard combats, the antimoderns never managed to innovate, occupying the minor role that was reserved for them. It cannot even be said in their favour that they put the brakes on the moderns' frenzy – those moderns for whom the antimoderns were always, in effect, the best of stooges.

The balance sheet of this examination is not too unfavourable. We can keep the Enlightenment without modernity, provided that we reintegrate the objects of the sciences and technologies into the Constitution, as quasi-objects among many others – objects whose genesis must no longer be clandestine, but must be followed through and through, from the hot events that spawned the objects to the progressive cool-down that transforms them into essences of Nature or Society.

Is it possible to draw up a Constitution that would allow us to recognize this work officially? We must do this, since old-style modernization can no longer absorb either other peoples or Nature; such, at least, is the conviction on which this essay is based. For its own good, the modern world can no longer extend itself without becoming once again what it has never ceased to be in practice – that is, a nonmodern world like all the others. This fraternity is essential if we are to absorb the two sets of entities that revolutionary modernization left behind: the natural crowds that we no longer master, the human multitudes that no one dominates any longer. Modern temporality gave the impression of continuous acceleration by relegating ever-larger masses of humans and nonhumans together to the void of the past.

Irreversibility has changed sides. If there is one thing we can no longer get rid of, it is those natures and multitudes, both equally global. The political task starts up again, at a new cost. It has been necessary to modify the fabric of our collectives from top to bottom in order to absorb the citizen of the eighteenth century and the worker of the nineteenth. We shall have to transform ourselves just as thoroughly in order to make room, today, for the nonhumans created by science and technology.

5.3 Humanism Redistributed

Before we can amend the Constitution, we first have to relocate the human, to which humanism does not render sufficient justice. Here are some of the magnificent figures that the moderns have been able to depict and preserve: the free agent, the citizen builder of the Leviathan, the distressing visage of the human person, the other of a relationship, consciousness, the *cogito*, the hermeneut, the inner self, the thee and thou of dialogue, presence to oneself, intersubjectivity. But all these figures remain asymmetrical, for they are the counterpart of the object of the sciences – an object that remains orphaned, abandoned in the hands of those whom epistemologists, like sociologists, deem reductive, objective, rational. Where are the Mouniers of machines, the Lévinases of animals, the Ricoeurs of facts? Yet the human, as we now understand, cannot be grasped and saved unless that other part of itself, the share of things, is restored to it. So long as humanism is constructed through contrast with the object that has been abandoned to epistemology, neither the human nor the nonhuman can be understood.

   Where are we to situate the human? A historical succession of quasi-objects, quasi-subjects, it is impossible to define the human by an essence, as we have known for a long time. Its history and its anthropology are too diverse for it to be pinned down once and for all. But Sartre's clever move, defining it as a free existence uprooting itself from a nature devoid of significance, is obviously not one we can make, since we have invested all quasi-objects with action, will, meaning, and even speech. There is no longer a practico-inert where the pure liberty of human existence can get bogged down. To oppose it to the crossed-out God (or, conversely, to reconcile it with Him) is equally impossible, since it is by virtue of their common opposition to Nature that the modern Constitution has defined all three. Must the human be steeped in Nature, then? But if we were to go looking for specific results of specific scientific disciplines that would clothe this robot animated with neurons, impulses, selfish genes, elementary needs and economic calculations, we would never get beyond monsters and masks. The sciences multiply new definitions of humans without managing to displace the former ones, reduce them to any homogeneous one, or unify them. They add reality; they do not subtract it. The hybrids that they invent in the laboratory are still more exotic than those they claim to break down.

Must we solemnly announce the death of man and dissolve him in the play of language, an evanescent reflection of inhuman structures that would escape all understanding? No, since we are no more in Discourse than we are in Nature. In any event, nothing is sufficiently inhuman to dissolve human beings in it and announce their death. Their will, their actions, their words are too abundant. Will we have to avoid the question by making the human something transcendental that would distance us for ever from mere nature? This would amount to falling back on just one of the poles of the modern Constitution. Will we have to use force to extend some provisional and particular definition inscribed in the rights of man or the preambles of constitutions? This would amount to tracing out once again the two Great Divides, and believing in modernization.

If the human does not possess a stable form, it is not formless for all that. If, instead of attaching it to one constitutional pole or the other, we move it closer to the middle, it becomes the mediator and even the intersection of the two. The human is not a constitutional pole to be opposed to that of the nonhuman. The two expressions 'humans' and 'nonhumans' are belated results that no longer suffice to designate the other dimension. The scale of value consists not in shifting the definition of the human along the horizontal line that connects the Object pole to the Subject pole, but in sliding it along the vertical dimension that defines the nonmodern world. Reveal its work of mediation, and it will take on human form. Conceal it again, and we shall have to talk about inhumanity, even if it is draping itself in the Bill of Rights. The expression 'anthropomorphic' considerably underestimates our humanity. We should be talking about morphism. Morphism is the place where technomorphisms, zoomorphisms, phusimorphisms, ideomorphisms, theomorphisms, sociomorphisms, psychomorphisms, all come together. Their alliances and their exchanges, taken together, are what define the *anthropos*. A weaver of morphisms – isn't that enough of a definition? The closer the *anthropos* comes to this distribution, the more human it is. The farther away it moves, the more it takes on multiple forms in which its humanity quickly becomes indiscernible, even if its figures are those of the person, the individual or the self. By seeking to isolate its form from those it churns together, one does not defend humanism, one loses it.

How could the *anthropos* be threatened by machines? It has made them, it has put itself into them, it has divided up its own members among their members, it has built its own body with them. How could it be threatened by objects? They have all been quasi-subjects circulating within the collective they traced. It is made of them as much as they are made of it. It has defined itself by multiplying things. How could it be deceived by politics? Politics is its own making, in that it reconstructs the collective through continual controversies over representation that allow it to say, at every moment, what it is and what it wants. How could it be dimmed by religion? It is through religion that humans are linked to all their fellows, that they know themselves as persons. How could it be manipulated by the economy? Its provisional form cannot be assigned without the circulation of goods and obligations, without the continuous distribution of social goods that we concoct through the goodwill of things. *Ecce homo*: delegated, mediated, distributed, mandated, uttered. Where does the threat come from? From those who seek to reduce it to an essence and who – by scorning things, objects, machines and the social, by cutting off all delegations and senders – make humanism a fragile and precious thing at risk of being overwhelmed by Nature, Society, or God.

Modern humanists are reductionist because they seek to attribute action to a small number of powers, leaving the rest of the world with nothing but simple mute forces. It is true that by redistributing the action among all these mediators, we lose the reduced form of humanity, but we gain another form, which has to be called irreducible. The human is in the delegation itself, in the pass, in the sending, in the continuous exchange of forms. Of course it is not a thing, but things are not things either. Of course it is not a merchandise, but merchandise is not merchandise either. Of course it is not a machine, but anyone who has seen machines knows that they are scarcely mechanical. Of course it is not of this world, but this world is not of this world either. Of course it is not in God, but what relation is there between the God above and the God below? Humanism can maintain itself only by sharing itself with all these mandatees. Human nature is the set of its delegates and its representatives, its figures and its messengers. That symmetrical universal is worth at least as much as the moderns' doubly asymmetrical one. This new position, shifted in relation to the subject/society position, now needs to be underwritten by an amended Constitution.

5.4 The Nonmodern Constitution

In the course of this essay, I have simply reestablished symmetry between the two branches of government, that of things – called science and technology – and that of human beings. I have also shown why the separation of powers between the two branches, after allowing for the proliferation of hybrids, could no longer worthily represent this new third estate. A constitution is judged by the guarantees it offers. The moderns' Constitution – as we recall from Section 2.8 – included four guarantees that had meaning only when they were taken together but also kept strictly separate. The first one guaranteed Nature its transcendent dimension by making it distinct from the fabric of Society – thus contrary to the continuous connection between the natural order and the social order found among the premoderns. The second guaranteed Society its immanent dimension by rendering citizens totally free to reconstruct it artificially – as opposed to the continuous connection between the social order and the natural order that kept the premoderns from being able to modify the one without modifying the other. But as that double separation allowed in practice for the mobilization and construction of Nature (Nature having become immanent through mobilization and construction) – and, conversely, made it possible to make Society stable and durable (Society having become transcendent owing to the enrolment of ever more numerous nonhumans), a third guarantee assured the separation of powers, the two branches of government being kept in separate, watertight compartments: even though it is mobilizable and constructed, Nature will remain without relation to Society; Society, in turn, even though it is transcendent and rendered durable by the mediation of objects, will no longer have any relation to Nature. In other words, quasi-objects will be officially banished – should we say taboo? – and translation networks will go into hiding, offering to the work of purification a counterpart that will nevertheless continue to be followed and monitored – until the postmoderns obliterate it entirely. The fourth guarantee of the crossed-out God made it possible to stabilize this dualist and asymmetrical mechanism by ensuring a function of arbitration, but one without presence or power (see Section 2.9).

In order to sketch in the nonmodern Constitution, it suffices to take into account what the modern Constitution left out, and to sort out the guarantees we wish to keep. We have committed ourselves to providing representation for quasi-objects. It is the third guarantee of the modern Constitution that must therefore be suppressed, since that is the one that made the continuity of their analysis impossible. Nature and Society are not two distinct poles, but one and the same production of successive states of societies-natures, of collectives. The first guarantee of our new draft thus becomes the nonseparability of quasi-objects, quasi-subjects. Every concept, every institution, every practice that interferes with the continuous deployment of collectives and their experimentation with hybrids will be deemed dangerous, harmful, and – we may as well say it – immoral. The work of mediation becomes the very centre of the double power, natural and social. The networks come out of hiding. The Middle

Kingdom is represented. The third estate, which was nothing, becomes everything.

As I have suggested, however, we do not wish to become premoderns all over again. The nonseparability of natures and societies had the disadvantage of making experimentation on a large scale impossible, since every transformation of nature had to be in harmony with a social transformation, term for term, and vice versa. Now we seek to keep the moderns' major innovation: the separability of a nature that no one has constructed – transcendence – and the freedom of manœuvre of a society that is of our own making – immanence. Nevertheless, we do not seek to inherit the clandestineness of the inverse mechanism that makes it possible to construct Nature – immanence – and to stabilize Society durably – transcendence.

Can we retain the first two guarantees of the old Constitution without maintaining the now-visible duplicity of its third guarantee? Yes, although at first this looks like squaring the circle. Nature's transcendence, its objectivity, and Society's immanence, its subjectivity, stem from the work of mediation without depending on their separation, contrary to what the Constitution of the moderns claims. The work of producing a nature or producing a society stems from the durable and irreversible accomplishment of the common work of delegation and translation. At the end of the process, there is indeed a nature that we have not made, and a society that we are free to change; there are indeed indisputable scientific facts, and free citizens, but once they are viewed in a nonmodern light they become the double consequence of a practice that is now visible in its continuity, instead of being, as for the moderns, the remote and opposing causes of an invisible practice that contradicts them. The second guarantee of our new draft thus makes it possible to recover the first two guarantees of the modern Constitution but without separating them. All concepts, all institutions, all practices that interfere with the progressive objectivization of Nature – incorporation into a black box – and simultaneously the subjectivization of Society – freedom of manœuvre – will be deemed harmful, dangerous and, quite simply, immoral. Without this second guarantee, the networks liberated by the first would keep their wild and uncontrollable character. The moderns were not mistaken in seeking objective nonhumans and free societies. They were mistaken only in their certainty that that double production required an absolute distinction between the two terms and the continual repression of the work of mediation.

Historicity found no place in the modern Constitution because it was framed by the only three entities whose existence it recognized. Contingent history existed for humans alone, and revolution became the only way for the moderns to understand their past – as I have shown in

Section 3.8, above – by breaking totally with it. But time is not a smooth, homogeneous flow. If time depends on associations, associations do not depend on time. We are no longer going to be confronted with the argument of time that passes for ever based on a regrouping into a coherent set of elements that belong to all times and all ontologies. If we want to recover the capacity to sort that appears essential to our morality and defines the human, it is essential that no coherent temporal flow comes to limit our freedom of choice. The third guarantee, as important as the others, is that we can combine associations freely without ever confronting the choice between archaism and modernization, the local and the global, the cultural and the universal, the natural and the social. Freedom has moved away from the social pole it had occupied exclusively during the modern representation into the middle and lower zones, and becomes a capacity for sorting and recombining sociotechnological imbroglios. Every new call to revolution, any epistemological break, any Copernican upheaval, any claim that certain practices have become outdated for ever, will be deemed dangerous, or – what is still worse in the eyes of the moderns – outdated!

Modern Constitution	Nonmodern Constitution
1st guarantee: Nature is transcendent but mobilizable (immanent).	*1st guarantee:* nonseparability of the common production of societies and natures.
2nd guarantee: Society is immanent but it infinitely surpasses us (transcendent)	*2nd guarantee:* continuous following of the production of Nature, which is objective, and the production of Society, which is free. In the last analysis, there is indeed a transcendence of Nature and an immanence of Society, but the two are not separated.
3rd guarantee: Nature and Society are totally distinct, and the work of purification bears no relation to the work of mediation.	*3rd guarantee:* freedom is redefined as a capacity to sort the combinations of hybrids that no longer depend on a homogeneous temporal flow.
4th guarantee: the crossed-out God is totally absent but ensures arbitration between the two branches of government.	*4th guarantee:* the production of hybrids, by becoming explicit and collective, becomes the object of an enlarged democracy that regulates or slows down its cadence.

Figure 5.2 Modern/nonmodern constitutions

But if I am right in my interpretation of the modern Constitution, if it has really allowed the development of collectives while officially forbidding what it permits in practice, how could we continue to develop quasi-objects, now that we have made their practice visible and official? By offering guarantees to replace the previous ones, are we not making impossible this double language, and thus the growth of collectives? That is precisely what we want to do. This slowing down, this moderation, this regulation, is what we expect from our morality. The fourth guarantee – perhaps the most important – is to replace the clandestine proliferation of hybrids by their regulated and commonly-agreed-upon production. It is time, perhaps, to speak of democracy again, but of a democracy extended to things themselves. We are not going to be caught by Archimedes' coup again.

Do we need to add that the crossed-out God, in this new Constitution, turns out to be liberated from the unworthy position to which He had been relegated? The question of God is reopened, and the nonmoderns no longer have to try to generalize the improbable metaphysics of the moderns that forced them to believe in belief.

5.5 The Parliament of Things

We want the meticulous sorting of quasi-objects to become possible – no longer unofficially and under the table, but officially and in broad daylight. In this desire to bring to light, to incorporate into language, to make public, we continue to identify with the intuition of the Enlightenment. But this intuition has never had the anthropology it deserved. It has divided up the human and the nonhuman and believed that the others, rendered premoderns by contrast, were not supposed to do the same thing. While it was necessary, perhaps, to increase mobilization and lengthen some networks, this division has now become superfluous, immoral, and – to put it bluntly – anti-Constitutional! We have been modern. Very well. We can no longer be modern in the same way. When we amend the Constitution, we continue to believe in the sciences, but instead of taking in their objectivity, their truth, their coldness, their extraterritoriality – qualities they have never had, except after the arbitrary withdrawal of epistemology – we retain what has always been most interesting about them: their daring, their experimentation, their uncertainty, their warmth, their incongruous blend of hybrids, their crazy ability to reconstitute the social bond. We take away from them only the mystery of their birth and the danger their clandestineness posed to democracy.

Yes, we are indeed the heirs of the Enlightenment, whose asymmetrical rationality is just not broad enough for us. Boyle's descendants had defined a parliament of mutes, the laboratory, where scientists, mere intermediaries, spoke all by themselves in the name of things. What did these representatives say? Nothing but what the things would have said on their own, had they only been able to speak. Outside the laboratory,

Hobbes's descendants had defined the Republic in which naked citizens, unable to speak all at once, arranged to have themselves represented by one of their number, the Sovereign, a simple intermediary and spokesperson. What did this representative say? Nothing but what the citizens would have said had they all been able to speak at the same time. But a doubt about the quality of that double translation crept in straight away. What if the scientists were talking about themselves instead of about things? And if the Sovereign were pursuing his own interests instead of reciting the script written for him by his constituents? In the first case, we would lose Nature and fall back into human disputes; in the second, we would fall back into the State of Nature and into the war of every man against every man. By defining a total separation between the scientific and political representations, the double translation-betrayal became possible. We shall never know whether scientists translate or betray. We shall never know whether representatives betray or translate.

During the modern period, the critics will continue to sustain themselves on that double doubt and the impossibility of ever putting an end to it. Modernism consisted in choosing that arrangement, nevertheless, but in remaining constantly suspicious of its two types of representatives without combining them into a single problem. Epistemologists wondered about scientific realism and the faithfulness of science to things; political scientists wondered about the representative system and the relative faithfulness of elected officials and spokespersons. All had in common a hatred of intermediaries and a desire for an immediate world, emptied of its mediators. All thought that this was the price of faithful representation, without ever understanding that the solution to their problem lay in the other branch of government.

In the course of this essay, I have shown what happened once science studies re-examined such a division of labour. I have shown how fast the modern Constitution broke down, since it no longer permitted the construction of a common dwelling to shelter the societies-natures that the moderns have bequeathed us. There are not two problems of representation, just one. There are not two branches, only one, whose products can be distinguished only late in the game, and after being examined together. Scientists appear to be betraying external reality only because they are constructing their societies and their natures at the same time. The Sovereign appears to be betraying his constituents only because he is churning together both citizens and the enormous mass of nonhumans that allow the Leviathan to hold up. Suspicion about scientific representation stemmed only from the belief that without social pollution Nature would be immediately accessible. 'Eliminate the social and you will finally have a faithful representation,' said some. 'Eliminate objects and you will finally have a faithful representation,' declared others. Their whole debate arose from the division of powers enforced by the modern Constitution.

Let us again take up the two representations and the double doubt about the faithfulness of the representatives, and we shall have defined the Parliament of Things. In its confines, the continuity of the collective is reconfigured. There are no more naked truths, but there are no more naked citizens, either. The mediators have the whole space to themselves. The Enlightenment has a dwelling-place at last. Natures are present, but with their representatives, scientists who speak in their name. Societies are present, but with the objects that have been serving as their ballast from time immemorial. Let one of the representatives talk, for instance, about the ozone hole, another represent the Monsanto chemical industry, a third the workers of the same chemical industry, another the voters of New Hampshire, a fifth the meteorology of the polar regions; let still another speak in the name of the State; what does it matter, so long as they are all talking about the same thing, about a quasi-object they have all created, the object-discourse-nature-society whose new properties astound us all and whose network extends from my refrigerator to the Antarctic by way of chemistry, law, the State, the economy, and satellites. The imbroglios and networks that had no place now have the whole place to themselves. They are the ones that have to be represented; it is around them that the Parliament of Things gathers henceforth. 'It was the stone rejected by the builders that became the keystone' (Mark 12:10).

However, we do not have to create this Parliament out of whole cloth, by calling for yet another revolution. We simply have to ratify what we have always done, provided that we reconsider our past, provided that we understand retrospectively to what extent we have never been modern, and provided that we rejoin the two halves of the symbol broken by Hobbes and Boyle as a sign of recognition. Half of our politics is constructed in science and technology. The other half of Nature is constructed in societies. Let us patch the two back together, and the political task can begin again.

Is it asking too little simply to ratify in public what is already happening? Should we not strive for more glamorous and more revolutionary programmes of action, rather than underlining what is already dimly discernible in the shared practices of scientists, politicians, consumers, industrialists and citizens when they engage in the numerous sociotechnological controversies we read about daily in our newspapers? As we have been discovering throughout this essay, the official representation is effective; that representation is what allowed, under the old Constitution, the exploration and proliferation of hybrids. Modernism was not an illusion, but an active performing. If we could draft a new

Constitution, we would, similarly, profoundly alter the course of quasi-objects. Another Constitution will be just as effective, but it will produce different hybrids. Is that too much to expect of a change in representation that seems to depend only on the scrap of paper of a Constitution? It may well be; but there are times when new words are needed to convene a new assembly. The task of our predecessors was no less daunting when they invented rights to give to citizens or the integration of workers into the fabric of our societies. I have done my job as philosopher and constituent by gathering together the scattered themes of a comparative anthropology. Others will be able to convene the Parliament of Things.

We scarcely have much choice. If we do not change the common dwelling, we shall not absorb in it the other cultures that we can no longer dominate, and we shall be forever incapable of accommodating in it the environment that we can no longer control. Neither Nature nor the Others will become modern. It is up to us to change our ways of changing. Or else it will have been for naught that the Berlin Wall fell during the miraculous year 1989, offering us a unique practical lesson about the conjoined failure of socialism and naturalism.

BIBLIOGRAPHY

Althusser, Louis (1992), *L'avenir dure longtemps*, Paris: Stock.
Arendt, Hannah (1963), *Eichmann in Jerusalem: A report on the banality of evil*, New York: Viking Press.
Augé, Marc (1975), *Théorie des pouvoirs et idéologie*, Paris: Hermann.
Augé, Marc (1986), *Un ethnologue dans le métro*, Paris: Hachette.
Authier, Michel (1989), 'Archimède, le canon du savant', in *Éléments d'histoire des sciences*, ed. Michel Serres, pp. 101–28, Paris: Bordas.
Bachelard, Gaston (1967), *La Formation de l'esprit scientifique*, Paris: Vrin.
Barnes, Barry (1974), *Scientific Knowledge and Sociological Theory*, London: Routledge & Kegan Paul.
Barnes, Barry and Steven Shapin, eds. (1979), *Natural Order: Historical studies in scientific culture*, London: Sage.
Barthes, Roland ([1970] 1982), *The Empire of Signs*, New York: Hill & Wang.
Barthes, Roland ([1985] 1988), *The Semiotic Challenge*, New York: Hill & Wang.
Bastide, Françoise (In press) *Oeuvres de sémiotique des textes scientifiques*, Forthcoming, 1994.
Baudrillard, Jean (1992), *L'illusion de la fin, la grève des événements*, Paris: Galilée.
Bensaude-Vincent, Bernadette (1989), 'Lavoisier: une révolution scientifique', in *Éléments d'histoire des sciences*, ed. Michel Serres, pp. 363–86, Paris: Bordas.
Bijker, Wiebe E., Thomas P. Hughes and Trevor Pinch, eds. (1987), *The Social Construction of Technological Systems: New directions in the sociology and history of technology*, Cambridge, MA.: MIT Press.
Bloor, David ([1976] 1991), *Knowledge and Social Imagery* (2nd edn. with a new foreword), Chicago: University of Chicago Press.
Boltanski, Luc. (1990), *L'amour et la justice comme compétences*, Paris: A.-M. Métailié.
Boltanski, Luc and Laurent Thévenot (1991), *De la justification. Les économies de la grandeur*, Paris: Gallimard.

Bourdieu, Pierre and Loïc Wacquant (1992), *Réponses: Pour une anthropologie réflexive*, Paris: Le Seuil.

Bowker, Geoffrey and Bruno Latour (1987), 'A booming discipline short of discipline: social studies of science in France', *Social Studies of Science*, 17: 715–48.

Braudel, Fernand (1985), *The Perspective of the World: 15th to 18th century*, New York: Harper & Row.

Callon, Michel (1986), 'Some elements of a sociology of translation: domestication of the scallops and the fishermen of St Brieux Bay', in *Power, Action and Belief: A new sociology of Knowledge?*, ed. John Law, pp. 196–229, London: Routledge & Kegan Paul.

Callon, Michel, ed. (1989), *La science et ses réseaux: Genèse et circulation des faits scientifiques*, Paris: La Découverte.

Callon, Michel (1992), 'Techno-economic networks and irreversibility', in *A Sociology of Monsters: Essays on power, technology and domination*, ed. John Law, vol. 38, pp. 132–64. 38. London: Routledge Sociological Review Monograph.

Callon, Michel and Bruno Latour (1981), 'Unscrewing the Big Leviathans: how do actors macrostructure reality?', in *Advances in Social Theory and Methodology: Toward an integration of micro and macro sociologies*, ed. Karin Knorr and Aron Cicourel, pp. 277–303, London: Routledge.

Callon, Michel and Bruno Latour (1992), 'Don't throw the baby out with the Bath school! A reply to Collins and Yearley', in *Science as Practice and Culture*, ed. Andy Pickering, pp. 343–68. Chicago: University of Chicago Press.

Callon, Michel, John Law and Arie Rip, eds. (1986), *Mapping the Dynamics of Science and Technology*, London: Macmillan.

Cambrosio, Alberto, Camille Limoges and Denyse Pronovost (1990), 'Representing biotechnology: an ethnography of Quebec science policy', *Social Studies of Science* 20: 195–227.

Canguilhem, Georges ([1968] 1988), *Ideology and Rationality in the History of the Life Sciences*, transl. A. Goldhammer, Cambridge, MA: MIT Press.

Chandler, Alfred D. (1977) *The Visible Hand: The managerial revolution in American business*, Cambridge, MA: Harvard University Press.

Chandler, Alfred D. (1990), *Scale and Scope: The dynamics of industrial capitalism*, Cambridge, MA: Harvard University Press.

Chateauraynaud, Francis (1990), *Les affaires de faute professionnelle: Des figures de défaillance et des formes de jugement dans les situations de travail et devant les tribunaux*, doctoral thesis, Paris: École des Hautes Études en Sciences Sociales.

Clastres, Pierre (1974), *La société contre l'Etat*, Paris: Minuit.

Cohen, I. Bernard (1985), *Revolution in Science*, Cambridge, MA.: Harvard University Press.

Collins, Harry, M. (1985), *Changing Order: Replication and induction in scientific practice*, London and Los Angeles: Sage.

Collins, Harry M. and Steven Yearley (1992), 'Epistemological chicken', in *Science as Practice and Culture*, ed. Andy Pickering, pp. 301–26, Chicago: University of Chicago Press.

Collins, Harry M. and Trevor Pinch (1982), *Frames of Meaning: The social construction of extraordinary science*, London: Routledge & Kegan Paul.

Conklin, Harold (1983), *Ethnographic Atlas of the Ifugao: A study of environment*, New Haven, CT and London: Yale University Press.

Copans, J. and J. Jamin (1978), *Aux origines de l'anthropologie française*, Paris: Le Sycomore.

Cunningham, Andrew and Perry Williams, eds. (1992) *The Laboratory Revolution in Medicine*, Cambridge: Cambridge University Press.

Cussins, Adrian (1992). 'Content, embodiment and objectivity: the theory of cognitive trails,' *Mind*, 104.404: 651–88.

Dagognet, François (1989), *Éloge de l'objet: Pour une philosophie de la marchandise*, Paris: Vrin.

Deleuze, Gilles (1968), *Différence et répétition*, Paris: Presses Universitaires de France.

Deleuze, Gilles and Félix Guattari ([1972] 1983), *Anti-Oedipus: Capitalism and schizophrenia*, Minneapolis: University of Minnesota Press.

Descola, Philippe ([1986] 1993), *In the Society of Nature, Native Cosmology in Amazonia*, Cambridge: Cambridge University Press.

Desrosières, Alain (1990), 'How to make things which hold together: social science, statistics and the state', in *Discourses on Society*, P. Wagner, B. Wittcocq and R. Whittley, eds., Dordrecht: Kluwer Academic Publishers, pp. 195–218.

Douglas, Mary (1983), *Risk and Culture: An essay in the selection of technical and environmental dangers*, Berkeley: University of California Press.

Durkheim, Emile ([1915] 1965), *The Elementary Forms of the Religious Life*, New York: Free Press.

Durkheim, Emile and Marcel Mauss ([1903] 1967), *Primitive Classifications*, Chicago: University of Chicago Press.

Eco, Umberto (1979), *The Role of the Reader: Explorations in the semiotics of texts*, London: Hutchinson.

Eisenstein, Elizabeth (1979), *The Printing Press as an Agent of Change*, Cambridge: Cambridge University Press.

Ellul, Jacques (1967), *Technological Society*, New York: Random House.

Fabian, Johannes (1983), *Time and the other: How anthropology makes its object*, New York: Columbia University Press.

Favret-Saada, Jeanne (1980), *Deadly Words: Witchcraft in the bocage*, trans. Catherine Cullen, Cambridge: Cambridge University Press.

Funkenstein, A. (1986), *Theology and the Scientific Imagination from the Middle Ages*, Princeton: Princeton University Press.

Furet, François ([1978] 1981), *Interpreting the French Revolution*, transl. Elborg Forsher, Cambridge: Cambridge University Press.

Garfinkel, Harry (1967), *Studies in Ethnomethodology*, Englewood Cliffs, NJ: Prentice Hall.

Geertz, Clifford (1971), *The Interpretation of Cultures: Selected essays*, New York: Basic Books.

Girard, René (1983), 'La danse de Salomé', in *L'auto-organisation de la physique au politique*, ed. Paul Dumouchel and Jean-Pierre Dupuy, pp. 336–52, Paris: Le Seuil.

Girard, René ([1978] 1987), *Things Hidden Since the Foundation of the World*, Stanford, CA: Stanford University Press.

Girard, René (1989), *The Scapegoat*, Baltimore, MD: Johns Hopkins University Press.

Goody, Jack (1977), *The Domestication of the Savage Mind*, Cambridge: Cambridge University Press.

Goody, Jack (1986), *The Logic of Writing and the Organization of Society*, Cambridge: Cambridge University Press.

Greimas, Algirdas Julien (1976), *On Meaning: Selected writings in semiotic theory*, Minneapolis: University of Minnesota Press.

Greimas, A.J. and J. Courtès, eds. (1982), *Semiotics and Language: An analytical dictionary*, Bloomington: Indiana University Press.

Habermas, Jürgen ([1981] 1989), *The Theory of Communicative Action*, Boston, MA: Beacon Press.

Habermas, Jürgen ([1985] 1987), *The Philosophical Discourse of Modernity: Twelve lectures*, transl. Frederick Lawrence, Cambridge, MA: MIT Press.

Hacking, Ian (1983), *Representing and Intervening*, Cambridge: Cambridge University Press.

Haraway, Donna (1989), *Primate Visions: Gender, race and nature in the world*, London: Routledge & Kegan Paul.

Haraway, Donna (1991), *Simians, Cyborgs, and Women: The reinvention of nature*, New York: Chapman & Hall.

Haudricourt, A.G. (1962), 'Domestication des animaux, culture des plantes et traitement d'autrui', *L'Homme* 2: 40–50.

Heidegger, Martin (1977a), 'Letter on Humanism', in *Basic Writings*, ed. David Farrell Krell, pp. 189–242, New York: Harper & Row.

Heidegger, Martin (1977b), *The Question Concerning Technology and Other Essays*, New York: Harper Torch Books.

Hennion, Antoine (1991), 'La médiation musicale', doctoral thesis, Paris: École des Hautes Études en Sciences Sociales.

Hobbes, Thomas ([1914] 1947), *Leviathan, or the Matter, Forme and Power of a Commonwealth Ecclesiastical and Civil*, London: J. M. Dent.

Hollis, Martin and Stephen Lukes, eds. (1982), *Rationality and Relativism*, Oxford: Blackwell.

Horton, Robin (1967), 'African traditional thought and Western science,' *Africa* 37: 50–71, 155–87.

Horton, Robin (1982), 'Tradition and modernity revisited' in *Rationality and Relativism*, ed. Martin Hollis and Stephen Lukes, pp. 201–60, Oxford: Blackwell.

Hughes, Thomas P. (1983), *Networks of Power: Electric supply systems in the US., England and Germany, 1880–1930*, Baltimore, MD.: Johns Hopkins University Press.

Hull, David L. (1988), *Science as a Process: An evolutionary account of the social and conceptual development of science*, Chicago: University of Chicago Press.

Hutcheon, Linda (1989), *The Politics of Postmodernism*, London: Routledge.

Hutchins, Edward (1980), *Culture and Inference. A Trobriand case study*, Cambridge, MA.: Harvard University Press.

Jameson, Frederic (1991), *Postmodernism or the Cultural Logic of Late Capitalism*, New Brunswick: Duke University Press.

Jonsen, Albert R. and Stephen Toulmin (1988), *The Abuse of Casuistry. A history of moral reasoning*, Berkeley: University of California Press.

Kidder, Tracy (1981), *The Soul of a New Machine*, London: Allen Lane.

Knorr, Karin (1981), *The Manufacture of Knowledge: An essay on the constructivist and contextual nature of science*, Oxford: Pergamon Press.

Knorr-Cetina, Karin (1992) 'The couch, the cathedral and the laboratory: on the relationships between experiment and laboratory in science', in *Science as Practice and Culture*, ed. Andrew Pickering, pp. 113–38, Chicago: University of Chicago Press.

Lagrange, Pierre (1990), 'Enquête sur les soucoupes volantes', *Terrain* 14: 76–91.

Latour, Bruno (1977), 'La répétition de Charles Péguy', in *Péguy écrivain. Colloque du centenaire*, ed. Centre Charles Péguy, pp. 75–100, Paris: Klincksieck.

Latour, Bruno (1983), 'Give me a laboratory and I will raise the world', in *Science Observed*, ed. Karin Knorr-Cetina and Michael Mulkay, pp. 141–70, London: Sage.

Latour, Bruno (1987), *Science In Action: How to follow scientists and engineers through society*, Cambridge, MA.: Harvard University Press.

Latour, Bruno (1988a), *Irreductions. Part II of The Pasteurization of France*, Cambridge, MA.: Harvard University Press.

Latour, Bruno (1988b), *The Pasteurization of France*, Cambridge, MA: Harvard University Press.

Latour, Bruno (1988c), 'The prince for machines as well as for machinations', in *Technology and Social Change*, ed. Brian Elliott, pp. 20–43, Edinburgh: Edinburgh University Press.

Latour, Bruno, (1988d), 'A relativist account of Einstein's relativity', *Social Studies of Science* 18: 3–44.

Latour, Bruno (1990a), 'Drawing things together', in *Representation in Scientific Practice*, ed. Michael Lynch and Steve Woolgar, pp. 19–68, Cambridge, MA.: MIT Press.

Latour, Bruno (1990b), 'The force and reason of experiment', in *Experimental Inquiries: Historical, philosophical and social studies of experimentation in science*, ed. Homer Le Grand, pp. 49–80, Dordrecht: Kluwer Academic Publishers.

Latour, Bruno (1992a), *Aramis, ou l'amour des techniques*, Paris: La Découverte.

Latour, Bruno (1992b), 'One more turn after the social turn: easing science studies into the non-modern world', in *The Social Dimensions of Science*, ed. Ernan McMullin, pp. 272–92, Notre Dame: University of Notre Dame Press.

Latour, Bruno and Jocelyn De Noblet, eds. (1985), *Les "Vues" de l'esprit. Visualisation et Connaissance Scientifique*, Paris: Culture Technique.

Latour, Bruno and Steve Woolgar ([1979] 1986), *Laboratory Life: The construction of scientific facts* (2nd edn with a new postword), Princeton, NJ: Princeton University Press.

Law, John (1986), 'On the methods of long-distance control vessels navigation and the Portuguese route to India', in *Power, Action and Belief: A new sociology of knowledge?*, ed. John Law, pp. 234–63, London: Routledge & Kegan Paul.

Law, John, ed. (1992), *A Sociology of Monsters: Essays on power, technology and domination*, vol. 38, London: Routledge Sociological Review Monograph.

Law, John and Gordon Fyfe, eds. (1988), *Picturing Power: Visual depictions and social relations*, London: Routledge.

Lévi-Strauss, Claude ([1952] 1987), *Race and History*, Paris: UNESCO.

Lévi-Strauss, Claude ([1962] 1966), *The Savage Mind*, Chicago: University of Chicago Press.

Lévy, Pierre (1990), *Les technologies de l'intelligence: L'avenir de la pensée à l'ère informatique*, Paris: La Découverte.

Lynch, Michael and Steve Woolgar, eds. (1990), *Representation in Scientific Practice*, Cambridge, MA.: MIT Press.

Lyotard, Jean-François (1979), *The Postmodern Condition: A report on knowledge*, Minneapolis: University of Minnesota Press.

Lyotard, Jean-François (15 April 1988), 'Dialogue pour un temps de crise (interview collective)', *Le Monde*, p. xxxviii.

MacKenzie Donald A. (1981), *Statistics in Britain. 1865–1930*, Edinburgh: The Edinburgh University Press.

MacKenzie, Donald A. (1990), *Inventing Accuracy: A historical sociology of nuclear missile guidance systems*, Cambridge, MA.: MIT Press.

Mauss, Marcel ([1923] 1967), *The Gift: Forms and functions of exchange in archaic societies* (with a foreword by E. Evans-Pritchard), New York: W.W. Norton.

Mayer, Arno (1982), *The Persistence of the Old Regime: Europe to the Great War*, transl. Jonathan Mandelbaum, New York: Pantheon.

Mayer, Arno (1988), *Why Did the Heavens not Darken? The 'Final Solution' in History*, New York: Pantheon.

Moscovici, Serge (1977), *Essai sur l'histoire humaine de la nature*, Paris: Flammarion.

Pavel, Thomas (1986), *Fictional Worlds*, Cambridge, MA: Harvard University Press.

Pavel, Thomas (1989), *The Feud of Language: A history of structuralist thought*, New York: Blackwell.

Péguy, Charles (1961a), 'Clio. Dialogue de l'histoire et de l'âme païenne', in *Oeuvres en prose*, pp. 93–309, Paris: Gallimard, Éditions de La Pléiade.

Péguy, Charles (1961b), *Oeuvres en Prose 1909–1914*, Paris: Gallimard, Éditions de la Pléiade.

Pickering, Andrew (1980), 'The role of interests in high-energy physics: the choice between charm and colour', *Sociology of the Sciences* 4: 107–38.

Pickering, Andrew, ed. (1992), *Science as Practice and Culture*, Chicago: University of Chicago Press.

Pinch, Trevor (1986), *Confronting Nature: The sociology of neutrino detection*, Dordrecht: Reidel.

Rogoff, Barbara and Jean Lave, eds (1984), *Everyday Cognition: Its development in social context*, Cambridge, MA.: Harvard University Press.

Schaffer, Simon (1988), 'Astronomers mark time: discipline and the personal equation', *Science In Context* 2,1: 115–45.

Schaffer, Simon (1991), 'A manufactory of OHMS: Victorian metrology and its instrumentation', in *Invisible Connections*, eds, S. Cozzes and R. Bud, pp. 25–54, Bellingham Washington State: Spi Press.

Serres, Michel (1974), *La Traduction (Hermès III)*, Paris: Minuit.

Serres, Michel (1987), *Statues*, Paris: François Bourin.

Serres, Michel (1989), 'Gnomon: les débuts de la géométrie en Grêce', in *Éléments d'histoire des sciences*, pp. 63–100, Paris: Bordas.

Serres, Michel (1991), *Le tiers instruit*, Paris: Bourin.

Serres, Michel and Bruno Latour (1992), *Éclaircissements: Cinq entretiens avec Bruno Latour*, Paris: Bourin.

Shapin, Steven (1990), ' "The Mind is its own Place": Science and Solitude in seventeenth-century England', *Science in Context*, 4, 1: 191–218.

Shapin, Steven (1992), 'History of science and its sociological reconstruction', *History of Science* 20: 157–211.

Shapin, Steven (1984), 'Pump and circumstance: Robert Boyle's literary technology', *Social Studies of Science* 14: pp. 481–520.

Shapin, Steven (1989), 'The invisible technician', *American Scientist* 77: 553–63.

Shapin, Steven and Simon Schaffer (1985), *Leviathan and the Air-Pump: Hobbes, Boyle and the experimental life*, Princeton, NJ: Princeton University Press.

Smith, Crosbie and Norton Wise (1989), *Energy and Empire: A biographical study of Lord Kelvin*, Cambridge: Cambridge University Press.

Stengers, Isabelle (1983), *États et processus*, doctoral thesis, Brussels: Université Libre de Bruxelles.

Stocking, G.W., (ed.). (1983), *Observers Observed. Essays on ethnographic fieldwork*, Madison: University of Wisconsin Press.

Stocking, G.W., ed. (1986), *Objects and Others: Essays on museums and material cultures*, Madison: University of Wisconsin Press.

Strum, Shirley and Bruno Latour (1987), 'The meanings of social: from baboons to humans', *Information sur les Sciences Sociales/Social Science Information* 26: 783–802.

Thévenot, Laurent (1989), 'Équilibre et rationalité dans un univers complexe', *Revue Économique* 2: 147–97.

Thévenot, Laurent (1990), 'L'action qui convient: Les formes de l'action', *Raison pratique* 1: 39–69.

Tile, Mary (1984), *Bachelard. Science and Objectivity*, Cambridge: Cambridge University Press.

Traweek, Sharon (1988), *Beam Times and Life Times: The world of high energy physicists*, Cambridge, MA.: Harvard University Press.

Trevor-Roper, Hugh (1983), 'The Highland tradition of Scotland', in *The Invention of Tradition*, ed. Eric Hobsbawm, pp. 15–41, Cambridge: Cambridge University Press.

Tuzin, Donald F. (1980), *The Voice of the Tambaran: Truth and illusion in the Iharita Arapesh religion*, Berkeley: University of California Press.

Vatimo, Gianni (1987), *La fin de la modernité: Nihilisme et herméneutique dans la culture postmoderne*, Paris: Le Seuil.

Warwick, Andrew (1992), 'Cambridge mathematics and Cavendish physics: Cunningham Campbell and Einstein's relativity 1905–1911'. Part 1: The uses of theory', *Studies in History and Philosophy of Science*, 23: 625–56.

Weber, Max ([1920] 1958), *The Protestant Ethic and the Spirit of Capitalism* (with an introduction by Antony Giddens), New York: Charles Scribner's Sons.

Wilson, Bryan R., ed. (1970), *Rationality*, Oxford: Blackwell.

Woolgar, Steve (1988), *Science: The very idea*, London: Tavistock.

Zimmerman, Michael E. (1990), *Heidegger's Confrontation with Modernity: Technology, politics and art*, Bloomington: Indiana University Press.

Zonabend, Françoise (1989), *La presqu'île au nucléaire*, Paris: Odile Jacob.

INDEX